C0-AKP-584

WITHDRAWN

St. Louis Community
College

Sphere, Spheroid and Projections for Surveyors

Aspects of modern land surveying

Series editor: J. R. Smith, ARICS

Electromagnetic Distance Measurement
C. D. Burnside

Fundamentals of Survey Measurement and Analysis
M. A. R. Cooper

Hydrography for the Surveyor and Engineer
A. E. Ingham

Mapping from Aerial Photographs
C. D. Burnside

J. E. JACKSON

Sphere, Spheroid and Projections for Surveyors

A HALSTED PRESS BOOK

JOHN WILEY & SONS Inc.
New York

First published in Great Britain 1980 by
Granada Publishing Limited – Technical Books Division
Frogmore, St Albans, Herts AL2 2NF
and
3 Upper James Street, London W1R 4BP

Published in the U.S.A. by
Halsted Press, a Division of
John Wiley & Sons, Inc., New York

Library of Congress Cataloguing in Publication Data
Jackson, John Eric
Sphere, spheroid and projections for surveyors. –
(Aspects of modern land surveying).
1. Surveying – Mathematics
2. Trigonometry, Spherical
I. Title II. Series
516′.24 TA549 80–82507

ISBN 0 470–27044–6

Phototypesetting by Parkway Group, London and Abingdon
Printed in Great Britain by William Clowes (Beccles) Limited

Contents

Foreword

The early Chinese appear to have realised that the Earth was some form of spherical shape rather than flat but knowledge of this was slow to reach the Western World where the credit is usually given to Pythagoras (570–497 BC). However it was Eratosthenes in 276 BC who made the first serious attempt at quantifying the parameters. Since then enormous strides have been taken in the refinements of these parameters and the methods of measurements have become very sophisticated – including the use of orbiting satellites. It remains a fact however that there is still no way in which such a shape can be represented truly in all its aspects on a flat sheet of paper. Hence the need for map projections.

The surveyor can be very accurate in his measurements to the stars or sun or in his distances and angles on the Earth's surface but to relate them correctly to one another there is often a need to use spherical trigonometry.

Thus it is that this book aims to bring together the various facets of the surveyor's and geographer's interests that involve spherical trigonometry and the geometry of the ellipse and spheroid. The author is a very experienced practitioner in these fields and brings a new approach to the treatment of the subject as seen by a professional surveyor. It should have appeal in a number of disciplines and be of assistance to both student and practitioner.

J. R. Smith, ARICS
Principal Lecturer
Portsmouth Polytechnic

Preface

The three parts of this book may be described briefly as:
I *Spherical trigonometry:* The development of formulas that may be useful to a surveyor in connection with astronomical observations, survey computations and mapping systems.
II *Spheroid geometry:* Formulas for computation of geodetic surveys on a spheroid of small eccentricity.
III *Map projections:* Methods for the precise representation of a spherical or spheroidal surface on to a plane. Problems in relating ground survey measurements to a plane coordinate system.

The main application of spherical trigonometry is in the treatment of astronomical observations which surveyors must sometimes make in order to apply correct processes of adjustment and computation to measurements made on the surface of the Earth. Spherical trigonometry has applications also in geography, cartography, navigation, geophysics and other disciplines.

Extensive survey systems for mapping purposes have always been reduced to two-dimensional geometry by representing them on a spheroidal surface, since this is the simplest form of surface that can be placed so as to be a sufficiently close fit to the somewhat irregular surface called the geoid, or mean sea level.

For the construction of maps, precise methods and formulas are required, to transform positions on the curved surface to a plane coordinate system. Such a transformation is a map projection. Many projection systems have been devised. The cartographer should have an extensive knowledge of these. National survey systems are expressed as plane coordinates of points. Land surveyors and civil engineers should know how to relate new survey measurements to such a system, taking account of the structure of the projection in which the coordinates have been calculated.

I am grateful to Professor Gordon Petrie of the University of Glasgow for making arrangements for the fair drawings, which were made by Mrs Hadden.

Cambridge J. E. Jackson

Part I Spherical trigonometry

CHAPTER 1

Angles only

In any situation where we are concerned with the relative directions of a number of lines radiating from a single origin, we may think of a sphere with its centre at that origin. Each radiating line will cut the sphere at a point; if one of the lines moves, its corresponding point will trace out a curve on the sphere. Thus the geometry on the sphere is an expression of the relative positions of the radiating lines. What is significant about this is that the lines are in three-dimensional space but, so long as the linear lengths of the lines are not to be considered, the geometry of the situation is represented on the two-dimensional surface of the sphere.

We may go further and suggest that in any problem concerned only with angles and directions, it should be possible to proceed without introducing any linear quantity, that is without stating a scale.

This is what spherical trigonometry is all about. Its most direct applications in surveying are in field astronomy. To the surveyor, a star in the sky is simply an indicator of a known direction; its distance from him is enormous and irrelevant. The surveyor wants to find latitude, longitude, bearing and these are all angles.

1.1 Celestial lines

In field astronomy the observer with a theodolite is concerned with three directions radiating from the centre of the instrument, point O in fig. 1.1. Direction OZ is the vertical or plumb-line, that is the direction fixed by gravity at the place where the instrument is set up. When the theodolite is correctly 'levelled' its mechanical axis of rotation coincides with this line. Another line through O is OS, the direction to a star on which the observer sets the telescope. The third direction of concern is OP, which is the line through O and parallel to the Earth's axis of rotation. We often see photographs that have been exposed for long periods at night and they show circular tracks of stars due to the Earth's rotation. The centre of these circles plane. By definition, the pole of the sky determines the earthly direction of

Although in reality the directions OP and OS are fixed in space, the observer who is going round with the Earth senses that OZ is fixed, and it is OS that apparently moves. The rate is just perceptible in the telescope of an ordinary theodolite when the star is close to one of the lines of the diaphragm.

1.2 True north

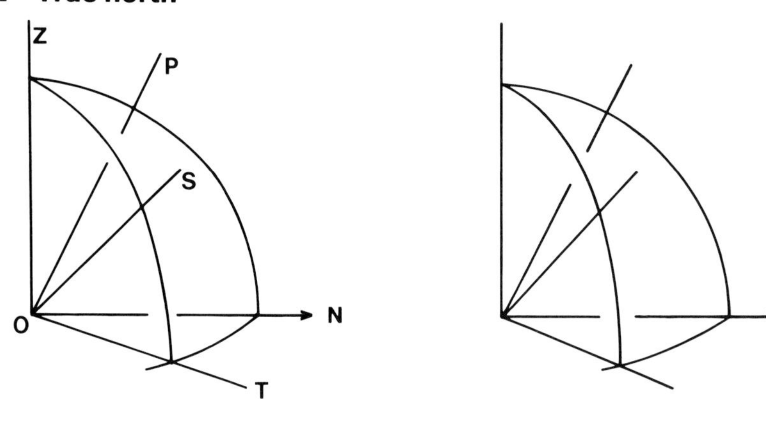

1.1 This diagram is a stereo-pair

Referring to fig. 1.1, think of the plane through O and perpendicular to OZ. This plane is horizontal at O, and any angle measured on the graduated horizontal plate of the theodolite can be seen as an angle about O in this plane. By definition, the pole of the sky determines the earthly direction of *true north*, or *true south* as the case may be. If there were a star exactly on the Earth's axis of rotation (there is not one), the sight line of the telescope could be set on it and then the telescope could be brought down to horizontal, when it would be pointing in the direction ON, true north for an observer in the northern hemisphere.

1.3 Azimuth

Similarly, the telescope could be pointed (momentarily) to a star and then brought down to the horizontal direction of the star, OT in the figure. The angle NOT is a horizontal angle reckoned clockwise from true north. In other words it is, or rather was, the bearing or *azimuth* of the star at the moment when the telescope was set on it.

1.4 Meridian

The plane of the three lines OZ, OP and ON is the meridian plane of the observer's instrument. One side of it is east and the other side of it is west. When a star is (momentarily) in this plane it is at *transit*.

1.5 Latitude

Astronomical latitude is usually defined as the angle made by OZ with the equatorial plane: it comes to the same thing to say that the angle ZOP is the co-latitude of the observer. It follows that the angle PON is the latitude and it is the elevation of the pole of the sky above horizontal. German writers usually refer to the astronomical latitude as *pole-height*.

1.6 Altitude

Another elevation or altitude to be seen on fig. 1.1 is angle TOS, the altitude of the star. This is the only angle in the situation that can be directly measured with a theodolite. In practice, of course, the angle read on the instrument must be corrected for atmospheric refraction. The complementary angle ZOS is called the *zenith distance* and on many modern theodolites the reading of the vertical circle gives this angle directly.

1.7 An astronomical operation

One of the practical jobs in field astronomy involves calculating the azimuth (angle NOT) of a star when a measurement of its altitude has been made. In this situation the two angles POZ (observer's co-latitude) and POS (*polar distance* of the star) are also known. Thus the relative directions of the three lines OZ, OP and OS, are known and it must be possible to calculate the azimuth. Clearly, the introduction of any linear dimension into the calculation should be unnecessary: the scale of fig. 1.1 is irrelevant.

1.8 Another kind of problem

As another example of a situation concerned only with angles, consider the simple contrived problem illustrated in fig. 1.2. A book is open so that the two opposite pages are at an angle of 50°. From a point H on the 'hinge' a line HK is drawn on one page at an angle of 45° to the hinge, and line HL is drawn on the other page at an angle of 30° to the hinge. What is the angle between HK and HL?

Evidently the answer does not depend on the physical size of the book, the lengths of HK and HL or on any other linear dimension. Nevertheless, one may be tempted to introduce a length, perhaps as a symbol or as some

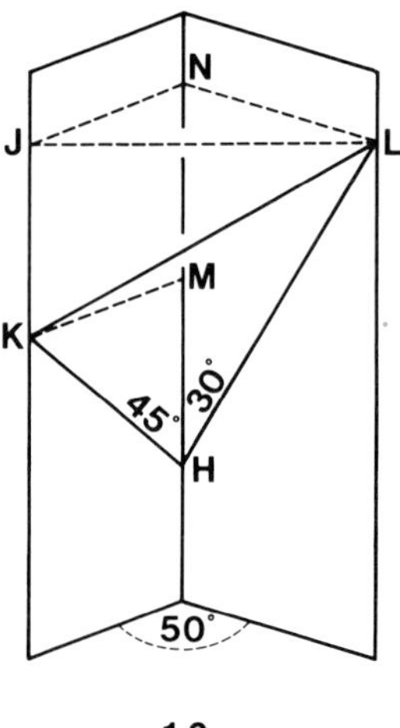

1.2

arbitrary numerical value, and proceed by calculating other lengths and eventually the required angle.
Suppose that the pages are 100 mm wide.
Then the length of HK is 100 cosec 45° = 141.421 mm
and the length of HL is 100 cosec 30° = 200.000 mm
KM and LN are perpendicular to the hinge and 100 mm long.
So HM is 100 cot 45° = 100.000 mm
and HN is 100 cot 30° = 173.205 mm.
Hence MN is 73.205 mm.

Now draw NJ perpendicular to the hinge. Then NJL is an isosceles triangle with sides 100 mm and angle 50°, so the length of JL is twice 100 sin 25°, that is 84.524 mm. JK is equal to NM. Finally, the length of KL can be calculated as the hypotenuse of the right-angled triangle LJK and the result is 111.818 mm. Now the lengths of the three sides of the plane triangle HKL are known and the angle at H can be found by a standard formula of plane trigonometry. The answer is 32° 53′ 55″.

CHAPTER 2

Spherical triangles

2.1 Six angles

In both the astronomical situation and the book problem, the essential geometry is a set of three concurrent lines. Consequently there are also three planes, each defined by a pair of the lines. And so, in addition to the three angles between pairs of lines there are three *dihedral angles* between pairs of the planes. An example is the angle JNL between the pages of the partly opened book. One of the dihedral angles is seen in fig. 1.1: it is the azimuth of the star, angle NOT, also seen at point Z where the two circular arcs meet, each of them perpendicular to OZ.

Precise relationships between the six angles must exist because, for instance, it is obvious that if the three angles between the pairs of lines are given, the geometry of the figure is determined and one would expect to be able to calculate the three dihedral angles. This is the analogue of the situation in plane geometry: when three sides of a triangle are given, the three angles can be calculated. Similarly, it is clear that if two of the angles between lines are given, as well as the dihedral angle between the two planes, as in the book problem, the geometry is also determined and one must be able to calculate the other three members of the sextet.

The study of the mathematical relationships among the set of six angles is the subject matter of spherical trigonometry.

2.2 Spherical triangle

A notation having no reference to the astronomical geometry will henceforth be used. Three lines, OA′, OB′ and OC′ are concurrent at O as in fig. 2.1. Consider a sphere with its centre at O. It is cut by the three lines at

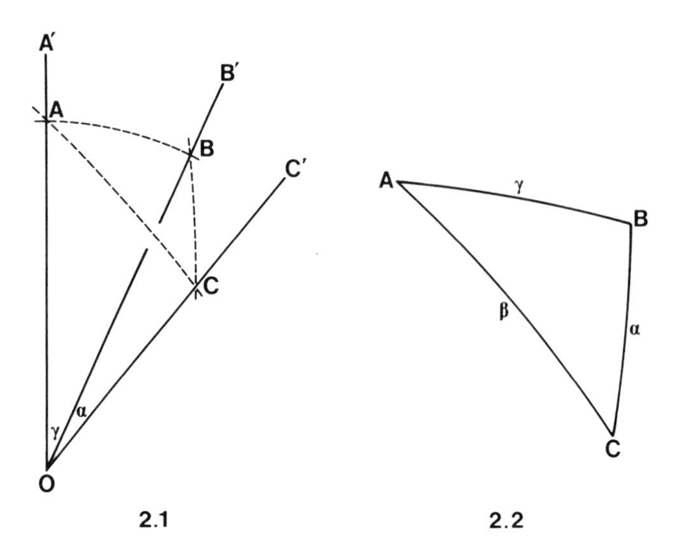

2.1 2.2

A, B and C. The plane of the two lines OA′ and OB′ will cut the surface of the sphere in a diametral circle partially shown by the curved broken line joining A and B on the diagram. A circle like this, which divides the sphere into two equal hemispheres, is called a *great circle.* Similarly there are great circles cut on the sphere by the planes B′OC′ and C′OA′.

This construction has generated a curvilinear triangle formed of three arcs of circles having the same radius as that of the sphere. These arcs meet the radii at right-angles at A, B and C and we now have visible representations of the three dihedral angles at the vertices of the curvilinear triangle. This is a *spherical triangle,* shown in isolation in fig. 2.2. The three circular arcs represent the angles BOC, COA and AOB which are here denoted by α, β and γ respectively. These are called the *sides* of the triangle. The angles at the vertices are denoted A, B and C respectively opposite to the sides α, β and γ.

2.3 Model triangle

A reader who has difficulty in visualising geometrical figures in three dimensions may find it very helpful to draw diagrams on a tennis ball or other suitable spherical object. This can make the geometry look trivially simple. Another way is to make a model spherical triangle by bending a sector of cardboard along two radii and joining the two edges marked OB as indicated in fig. 2.3. The curved edges form the spherical triangle, which in this case has sides 60°, 50° and 40°.

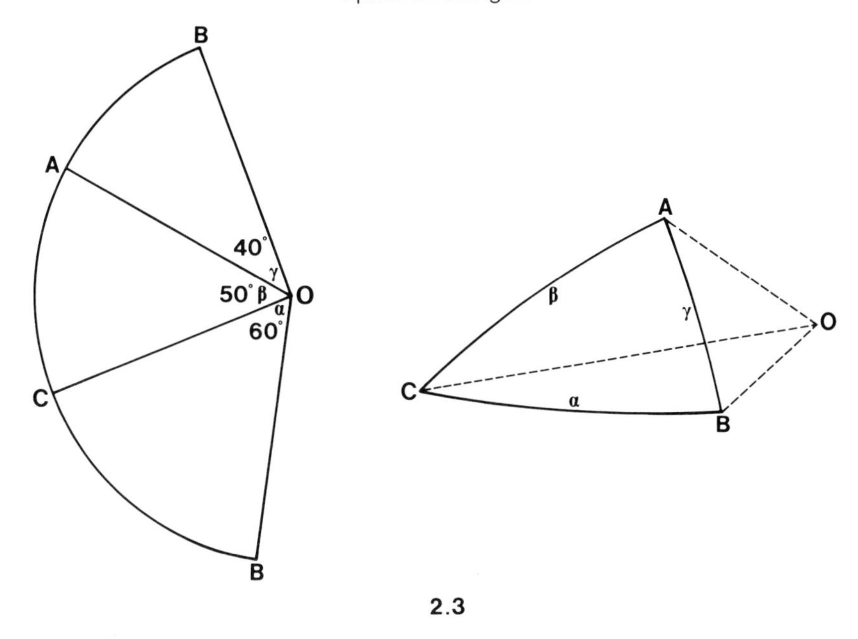

2.3

2.4 Spherical trigonometry

Spherical trigonometry is a collection of formulas expressing relations between trigonometrical functions of the six angles α, β, γ, A, B, C. Note that in the foregoing descriptions of the geometry, no mention of a linear dimension has been necessary. Of course, by use of a larger sphere centred at O one gets a larger triangle but the six angles it represents are the same. Where a three-dimensional problem to be solved is concerned only with angles, it is likely that formulas of spherical trigonometry will be applicable.

CHAPTER 3

Basic formulas

3.1 Sine formulas

Using the model in fig. 2.3, drop the perpendicular AK to the plane of OBC as shown in fig. 3.1, and perpendiculars KM to OB and KN to OC. Then the

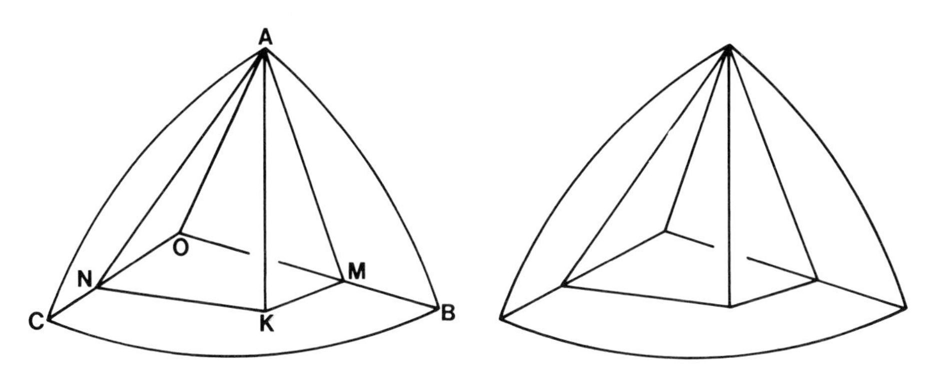

3.1 This diagram is a stereo-pair

triangles AKM and AKN are right-angled at K, the angle AMK is B and the angle ANK is C. Also, the triangle AMO is right-angled at M and the triangle ANO is right-angled at N. This may not be clearly seen on the diagram but it is quite obvious on the cardboard model. Angle AOM is γ and AON is β. Then if R is the radius of the sphere we see that:

$$\text{AM} = R\sin\gamma \qquad \text{AK} = R\sin\gamma\sin B$$
$$\text{AN} = R\sin\beta \qquad \text{AK} = R\sin\beta\sin C$$

On equating the two formulas for AK the R cancels out and the result is:

$$\frac{\sin\beta}{\sin B} = \frac{\sin\gamma}{\sin C}.$$

Obviously a similar construction with a perpendicular from one of the other vertices of the triangle would show that the above ratios are also equal to $\frac{\sin\alpha}{\sin A}$.

Thus we have the sine formulas:

$$\frac{\sin\alpha}{\sin A} = \frac{\sin\beta}{\sin B} = \frac{\sin\gamma}{\sin C}$$

very similar to the formulas for a plane triangle. Any pair of the sine formulas is a relationship between four of the angles associated with the spherical triangle.

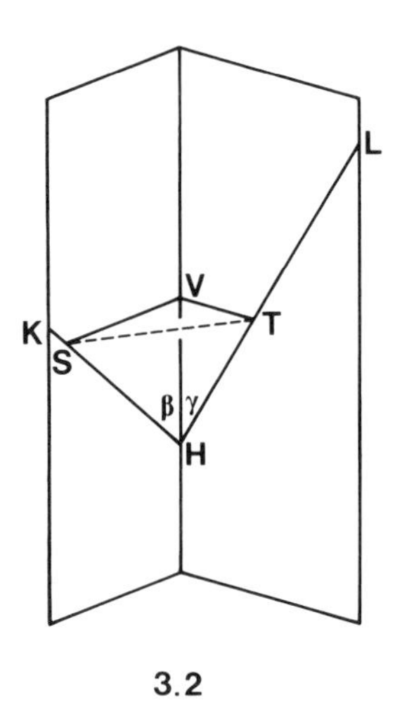

3.2

3.2 Cosine formulas

Another set of formulas is obtainable through a solution of the book problem in chapter 2 using the simpler geometrical construction shown in fig. 3.2. From a point V on the hinge, lines VS and VT are drawn perpendicular to the hinge to meet HK and HL. We now denote angle SHT as α, angle SHV as β, angle THV as γ, and the dihedral angle SVT as A. Let v be the length of HV. Then HS $= v\sec\beta$, HT $= v\sec\gamma$, VS $= v\tan\beta$ and VT $= v\tan\gamma$. Using a standard formula of plane trigonometry, the length of ST is calculated in both triangles SHT and SVT. The two expressions for $(\mathrm{ST})^2$ are:

$$v^2\sec^2\beta + v^2\sec^2\gamma - 2v^2\sec\beta\sec\gamma\cos\alpha$$

and

$$v^2\tan^2\beta + v^2\tan^2\gamma - 2v^2\tan\beta\tan\gamma\cos A.$$

Equating these terms, v^2 divides out and then the substitutions $\sec^2\beta = 1 + \tan^2\beta$ and $\sec^2\gamma = 1 + \tan^2\gamma$ are made. After some cancellations we get:

$$2\sec\beta\sec\gamma\cos\alpha = 2 + 2\tan\beta\tan\gamma\cos A.$$

Omitting the factor 2 and multiplying by $\cos\beta\cos\gamma$ gives the formula:

$$\cos\alpha = \cos\beta\cos\gamma + \sin\beta\sin\gamma\cos A.$$

This is a cosine formula. Note that it is a relationship between four of the 'elements' of the spherical triangle. In this case one side is expressed in terms of the other two sides and the angle included between them. It follows, obviously, that there are two other similar formulas, and the complete set of cosine formulas is:

$$\begin{aligned}\cos\alpha &= \cos\beta\cos\gamma + \sin\beta\sin\gamma\cos A\\ \cos\beta &= \cos\gamma\cos\alpha + \sin\gamma\sin\alpha\cos B\\ \cos\gamma &= \cos\alpha\cos\beta + \sin\alpha\sin\beta\cos C.\end{aligned}$$

So, for the book problem:

$$\begin{aligned}\cos \mathrm{SHT} &= \cos 30^\circ\cos 45^\circ + \sin 30^\circ\sin 45^\circ\cos 50^\circ\\ &= 0.839\,632\,2\\ \mathrm{SHT} &= 32^\circ\,53'\,55.3''\end{aligned}$$

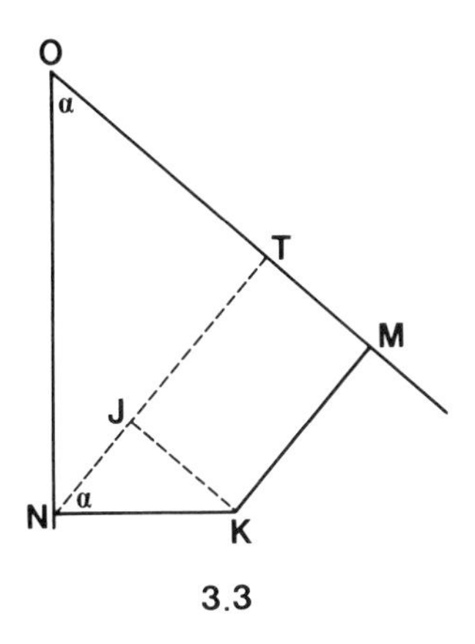

3.3

A cosine formula can be demonstrated also by further use of fig. 3.1 and the line NT drawn perpendicular to OB. The geometry on the plane OBC is shown in fig. 3.3. OM is $R\cos\gamma$. ON is $R\cos\beta$, so OT is $R\cos\beta\cos\alpha$. Angle TNK is obviously α. AN is $R\sin\beta$ so NK is $R\sin\beta\cos C$, hence TM = JK is $R\sin\beta\cos C\sin\alpha$. All these formulas are quite clear if the cardboard model is consulted. Using the fact that OM = OT + TM and omitting the multipler R gives:

$$\cos\gamma = \cos\alpha\cos\beta + \sin\alpha\sin\beta\cos C.$$

3.3 Inverted cosine formulas

By rearrangement, the cosine formulas become suitable for computation of the angles of the triangle when the sides are given. Thus:

$$\cos A = (\cos\alpha - \cos\beta\cos\gamma)/(\sin\beta\sin\gamma) \text{ etc.}$$

With the three sides 60°, 50° and 40°, the angles are:

$$\begin{aligned} A &= 89.116\,085° = 89°\ 06'\ 57.9'' \\ B &= 62.183\,505° = 62°\ 11'\ 00.6'' \\ C &= 47.913\,935° = 47°\ 54'\ 50.2'' \end{aligned}$$

and the three ratios of the sine formulas are each equal to 0.866 128, providing a check on the computations.

3.4 Spherical excess

It will be seen that the sum of the three angles above is 199° 12′ 48.7″. The extra value over 180° is called *spherical excess.* There is no fixed total for the angles of a spherical triangle; the excess depends on the ratio of the area of the triangle to the whole area of the sphere (*See* section 6.9.).

Think of a terrestrial globe, that is a spherical model representing the Earth. Take an arc of 90° along the equator and the two semi-meridians from the ends of this arc up to the North Pole. These three arcs form a spherical triangle in which all the sides are 90°, all the angles are 90° and the spherical excess is 90°.

3.5 Four-part formulas

Another set of useful formulas can be derived by taking a cosine formula and substituting another cosine formula into it, thus:

$$\cos\alpha = \cos\beta\,(\cos\alpha\cos\beta + \sin\alpha\sin\beta\cos C) + \sin\beta\sin\gamma\cos A.$$

Move the term $\cos\alpha\cos^2\beta$ from right to left and get:

$$\cos\alpha\sin^2\beta = \cos\beta\sin\alpha\sin\beta\cos C + \sin\beta\sin\gamma\cos A.$$

Then cancel factor $\sin\beta$ throughout, divide by $\sin\alpha$ and in the last term substitute $\sin C/\sin A$ for $\sin\gamma/\sin\alpha$ (using sine formulas) and the result is:

$$\sin\beta\cot\alpha = \cos\beta\cos C + \sin C\cot A.$$

This rearranges to:

$$\cos\beta\cos C = \sin\beta\cot\alpha - \sin C\cot A.$$

This is a four-part formula, so-called because the four angles appearing in it are consecutive elements round the triangle. Hence the full set has six of these formulas but they are not written out here because they can be made memorable by a verbal jingle:

A set of four consecutive elements has two inner (β and C above) and two outer (α and A) members. All the four-part formulas are described by:

cos (inner side) cos (inner angle)
equals sin (inner side) cot (outer side)
minus sin (inner angle) cot (outer angle).

In a four-part formula the outer elements appear only once, so its practical value is for calculating an outer element when three consecutive elements are given, for example:

$$\cot \alpha = (\cos \beta \cos C + \cot A \sin C)/\sin \beta.$$

3.6 Using formulas

In applications of spherical trigonometry the usual situation is that three of the six elements are known and it is required to calculate one or more of the other three. A practical requirement is that the element to be computed should appear only once in the formula so that an explicit expression for it is available in terms of the given three elements. This was alluded to in reference to the 'outer' elements in the four-part formulas; and in a cosine formula also, two of the elements appear only once. Any pair of the sine formulas gives one element explicitly in terms of three, but here there can be some ambiguity because the sine of an angle is the same as the sine of its supplement: sin 88° is the same as sin 92°. Anyway it is advisable not to calculate the sine of an angle that is likely to be close to 90°, nor the cosine of a very small angle, since any small uncertainty in the numerical value, such as a rounding-off error, will mean a large uncertainty in the deduced angle.

Evidently there is room for an intelligent choice of formula for practical computation work. Some more formulas and methods are described in the next chapter.

CHAPTER 4

More formulas

4.1 Right-angled triangle

When an angle of a spherical triangle is 90° the formulas are simpler, as is to be expected. For instance with $A = 90°$ the cosine formula for α is simply $\cos\alpha = \cos\beta\cos\gamma$. The right-angled triangle with sides 30° and 40° has hypotenuse $48.439237° = 48°\ 26'\ 21.3''$. Compare this with the plane triangle of sides 30 and 40 units and hypotenuse exactly 50 units. With sides of 30 and 40 angle minutes, the hypotenuse is 49.999 486 minutes or 2 999.97 seconds.

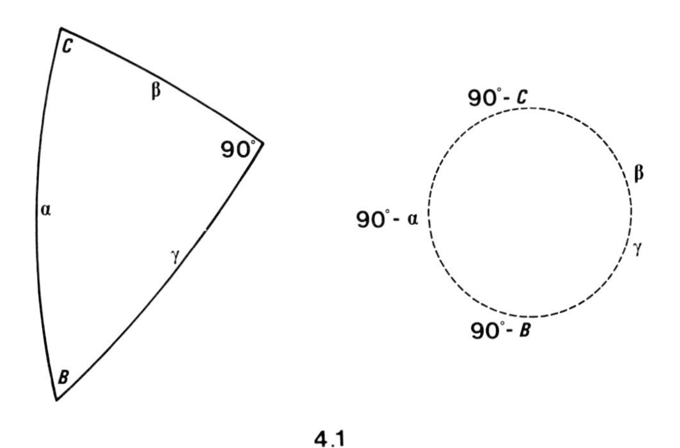

4.1

It has been found that the formulas applicable to right-angled spherical triangles can be obtained from an arrangement that is very easily memorised. As shown in fig. 4.1, five angles are set round a circle – the two sides containing the right angle and the complements of the other three elements. If we select any one of these five circular parts, as they are called, the following relations hold:

$$\begin{aligned}\text{sine (any part)} &= \text{product of tangents of (adjacent parts)}\\ &= \text{product of cosines of (opposite parts).}\end{aligned}$$

For instance:

$$\sin(90^\circ - B) = \tan\gamma\tan(90^\circ - \alpha) = \cos\beta\cos(90^\circ - C)$$

that is, $\cos B = \tan\gamma\cot\alpha \qquad = \cos\beta\sin C.$

Another formula is $\sin\beta = \sin\alpha\sin B$ which has obvious analogy with a plane triangle formula, and another is $\tan\beta = \tan\alpha\cos C$. Altogether ten simple formulas are obtainable.

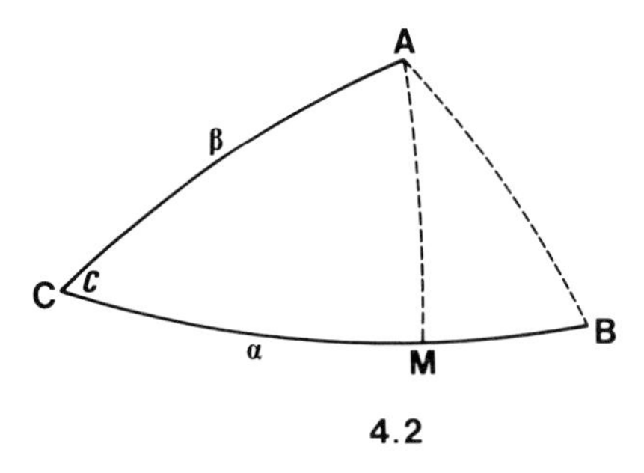

4.2

4.2 Using right-angled triangles

The use of right-angled triangles may provide alternative methods of solution of some problems in spherical trigonometry. Consider the case where two sides and the included angle are given, say α, β and C. As in fig. 4.2, two right-angled triangles are formed by drawing the perpendicular AM. In triangle ACM, β and C are known, so the sides AM and CM and the angle CAM can be calculated. In fact $\sin \text{AM} = \sin\beta\sin C$, $\tan CM = \tan\beta\cos C$ and $\cot \text{CAM} = \cos\beta\tan C$. Then $\text{MB} = \alpha - \text{CM}$, the triangle AMB is solved and the rest follows.

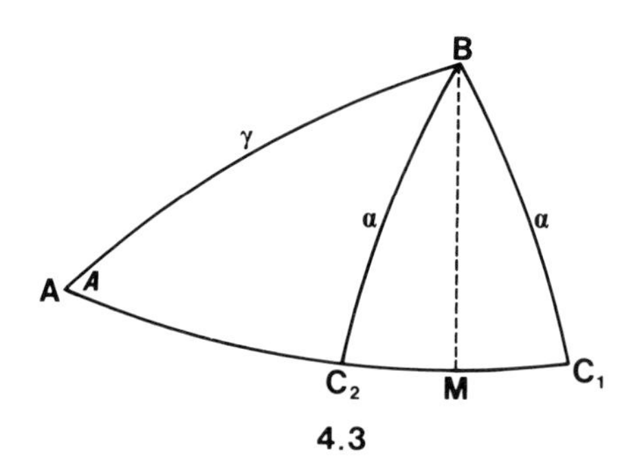

4.3

4.3 The ambiguous case

As in plane geometry, there will generally be two solutions if the given elements are two sides and the angle opposite one of them. Right-angled triangle formulas can be used here. Referring to fig. 4.3 it is seen that if A, α and γ are given, then both the triangles ABC_1 and ABC_2 are solutions. In many practical cases it is quite clear which is the appropriate one; otherwise some other test must be used to make the choice.

Using the sine formula $\sin C = \sin \gamma \sin A \operatorname{cosec} \alpha$ does not distinguish between C and $180° - C$. However, taking α with the acute angle C, we have $\tan MC_1 (= MC_2) = \tan \alpha \cos C$ and $\tan AM = \tan \gamma \cos A$, then the side β is $AM \pm MC_1$ as the case may be. Angle B is then obtainable from a sine formula.

The reader will see that some variations on the above procedure are possible. Angle B should be calculated by an inverted cosine formula if it is near 90°.

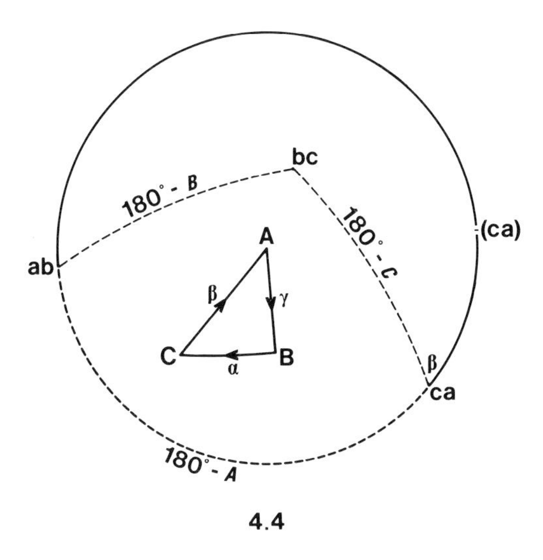

4.4

4.4 Polar triangle

Any great circle on a sphere has two poles just as the equator has on a terrestrial globe. In fig. 4.4, ABC is a spherical triangle and arrows are marked on the sides so that a particular pole of each circle is distinguished – the pole to the right of a traveller moving with the arrow. These three poles are marked bc, ca and ab. They also form a spherical triangle as shown with broken lines.

Now if the plane of the great circle CA is rotated anticlockwise about A until it coincides with the plane of the great circle AB, the pole of CA will move from ca to (ca) where it will then be exactly opposite to the point ab. Hence the side ab–ca of the polar triangle is $180° - A$. By similar arguments (which are not so easily followed on the plane diagram), the other sides of the polar triangle are $180° - B$ and $180° - C$.

The relation between the polar triangle and the original triangle is in fact reciprocal – each is a polar triangle of the other. Obviously A is a pole of the side ca–ab. The point B is 90° from bc because bc is a pole of the great circle BC; and B is also 90° from ab because ab is a pole of AB. Hence B is a pole of the side ab–bc. Similarly C is a pole of side bc–ca. Therefore the sides of ABC are the supplements of the angles of triangle bc–ca–ab.

The geometry described above is much clearer if the figure is drawn on a sphere! It will be seen, for instance, that if the great circle of which A is a pole is rotated about the diameter through ca, the point A will move along AC and will coincide with C when the rotation is equal to β; then ca–(ca) will lie along ca–bc. Thus the angle at ca must be β as marked and the angle of the polar triangle at ca is $180° - \beta$.

What all this wordy argument amounts to is that if a spherical triangle has sides α, β and γ, angles A, B, and C, then there exists a spherical triangle with *sides* $(180° - A)$, $(180° - B)$, $(180° - C)$ and *angles* $(180° - \alpha)$, $(180° - \beta)$, $(180° - \gamma)$. So we may take any formula of a spherical triangle and make the substitutions indicated above and the result will be a valid formula. For instance the cosine formula for α converts to:

$$\begin{aligned}\cos(180° - A) = {} & \cos(180° - B)\cos(180° - C) \\ & + \sin(180° - B)\sin(180° - C)\cos(180° - \alpha)\end{aligned}$$

which after tidying up is:

$$\cos A = -\cos B\cos C + \sin B\sin C\cos\alpha$$

and of course there are two others to make the set of three.

4.5 Sides from angles

By writing the above set of formulas as:

$$\cos\alpha = (\cos A + \cos B\cos C)/(\sin B\sin C) \text{ etc.,}$$

we see that if the angles of a spherical triangle are given the sides can be calculated. This is a difference from plane trigonometry. It corresponds with the fact that two spherical triangles cannot be *similar* in the sense of having the same angles but different sides.

4.6 Half-angle formulas

A cosine formula can be transformed with the aid of some standard trigonometrical identities as follows:

$$\begin{aligned}\cos\alpha &= \cos\beta\cos\gamma + \sin\beta\sin\gamma\,(1 - 2\sin^2\tfrac{1}{2}A) \\ &= \cos(\beta - \gamma) - 2\sin\beta\sin\gamma\sin^2\tfrac{1}{2}A.\end{aligned}$$

Hence, $$2\sin\beta\sin\gamma\sin^2\tfrac{1}{2}A = \cos(\beta-\gamma) - \cos\alpha$$
$$= 2\sin\tfrac{1}{2}(\beta-\gamma+\alpha)\sin\tfrac{1}{2}(\alpha-\beta+\gamma).$$

Now we write, $$\sigma = \tfrac{1}{2}(\alpha+\beta+\gamma).$$

Then, $$\tfrac{1}{2}(\beta-\gamma+\alpha) = (\sigma-\gamma)$$
and $$\tfrac{1}{2}(\alpha-\beta+\gamma) = (\sigma-\beta).$$

So we find that, $$\sin^2\tfrac{1}{2}A = \frac{\sin(\sigma-\beta)\sin(\sigma-\gamma)}{\sin\beta\sin\gamma}.$$

Going back to the cosine formula and writing $(2\cos^2\tfrac{1}{2}A - 1)$ instead of $\cos A$, leads similarly to:

$$\cos^2\tfrac{1}{2}A = \frac{\sin\sigma\sin(\sigma-\alpha)}{\sin\beta\sin\gamma}.$$

By division, $$\tan^2\tfrac{1}{2}A = \frac{\sin(\sigma-\beta)\sin(\sigma-\gamma)}{\sin\sigma\sin(\sigma-\alpha)}.$$

Obviously there are two similar formulas for $\tan^2\tfrac{1}{2}B$ and $\tan^2\tfrac{1}{2}C$.

The three angles can be verified by testing the sine formulas.

These half-angle formulas give the angles in terms of the sides and they are very convenient for computation by logarithms. Hence they were widely used in the days before electronic calculation.

Some similar formulas applicable in the two-sides-and-included-angle situation are:

$$\tan\tfrac{1}{2}(B+C) = \frac{\cos\tfrac{1}{2}(\beta-\gamma)}{\cos\tfrac{1}{2}(\beta+\gamma)}\cot\tfrac{1}{2}A$$

$$\tan\tfrac{1}{2}(B-C) = \frac{\sin\tfrac{1}{2}(\beta-\gamma)}{\sin\tfrac{1}{2}(\beta+\gamma)}\cot\tfrac{1}{2}A.$$

The sine formulas can be used to check B and C and to calculate the third side α.

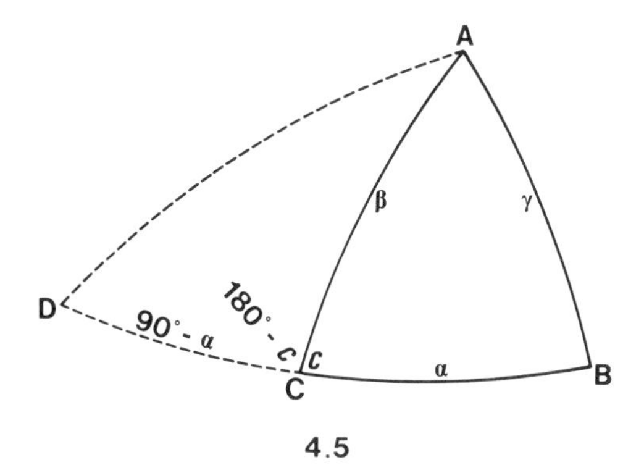

4.5

4.7 Another set of formulas

A set of formulas that can be useful in making mathematical transformations is obtainable by the method indicated in fig. 4.5. One side, in this case α, is extended to 90°, so that CD = 90° − α. Then by applying the cosine formula to AD in both the triangles ABD and ACD we find that:

$$\cos \mathrm{AD} = \sin \gamma \cos B = \sin \alpha \cos \beta - \cos \alpha \sin \beta \cos C.$$

Since each side can be extended either way a set of six formulas is available.

CHAPTER 5

Spherical astronomy

5.1 Astronomical triangle

The spherical triangle generated by the three lines OZ, OP and OS in fig. 1.1 has the sides shown in fig 5.1 in the usual astronomical notation. It so happens that the sides of the triangle are the complements of angles that are familiar in the astronomical context: ϕ is the latitude of the observer's station, δ is the declination of the star, and h is the altitude that can be measured with a theodolite (and must be corrected for atmospheric refraction). The sides themselves are *co-latitude, polar distance* and *zenith distance,* and these may of course be used with standard formulas when solving the triangle. Otherwise the formulas may be expressed in astronomical notation as is to be seen in most of the textbooks and manuals. Angle P is the

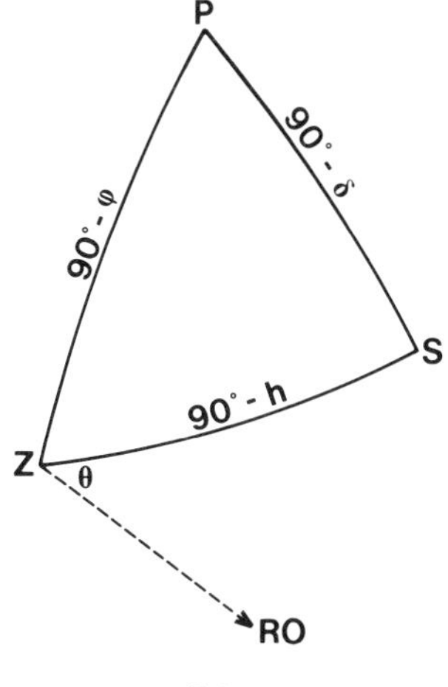

5.1

hour–angle and may also be seen denoted by H or T or t or other letters; angle Z is the *azimuth* or true bearing of the star sometimes denoted by A. It is now customary to reckon azimuth clockwise from north, so if the star is on the western side of the observer's meridian its azimuth is $360° - Z$. Angle S is called *parallactic angle* and is not a quantity required by the surveyor. If P is the North Pole and the declination of the star is south, then PS is $90° + \delta$.

The uses of the astronomical triangle in three basic operations of field astronomy are described briefly in the next three sections. In each of these situations the observer's latitude is assumed to be known and the side PS is found by reference to tables of *apparent places* such as the *Star Almanac for Land Surveyors* or the more precise *Apparent Places of Fundamental Stars*. To determine the triangle a third element is required: in two cases it is the side ZS and in the other case it is the angle at P.

It is common practice to change 'face' of the theodolite and make two settings on a star, so that the angles and chronometer readings that go into a computation are actually the means of pairs of recorded values.

5.2 Time by altitude

The altitude of a star is measured and simultaneously readings are taken of a chronometer or recorded on a chronograph. The hour–angle P is calculated. In terms of customary quantities the inverted cosine formula is:

$$\cos P = (\sin h - \sin \delta \sin \phi)/(\cos \delta \cos \phi).$$

Using the star's right ascension taken from the tables, the local time of the observation is calculated. The observation is thus a 'calibration' of the chronometer against the local time at the observer's meridian. By recording radio signals the chronometer's relationship to Universal Time can be found and thence the observer's longitude.

5.3 Azimuth by altitude

This operation involves measuring the altitude of a star and simultaneously its horizontal direction in relation to a signal on a fixed terrestrial reference mark (RO). The angle to be calculated is Z the star's azimuth:

$$\cos Z = (\sin \delta - \sin \phi \sin h)/(\cos \phi \cos h).$$

Adding or subtracting the measured horizontal angle (such as θ in fig. 5.1) gives what the surveyor wants – the azimuth of the fixed RO.

5.4 Azimuth by hour–angle

In this case the horizontal angle between the star and RO is measured and the chronometer reading at each setting on the star is precisely recorded.

The exact hour–angle of the star can then be calculated (assuming that the chronometer has been calibrated against local time) and the azimuth of the star is found from the appropriate four-part formula:

$$\cot Z = (\cos \phi \tan \delta - \sin \phi \cos P)/\sin P.$$

The measured horizontal angle is then applied.

An advantage of this method is that the altitude of the star is not required so there is no correction for refraction and stars at quite low altitudes can be used.

5.5 Elongation

The apparently circular tracks of stars round the Pole were mentioned in section 1.1. We see from this that there are stars that reach positions of maximum eastern or western direction relative to the meridian. At these positions the star is said to be at *elongation* and it is then (momentarily) moving vertically from the point of view of the observer. Setting the telescope on a star at or near elongation will thus be facilitated, especially for the accurate measurement of a horizontal angle between the star and a ground signal.

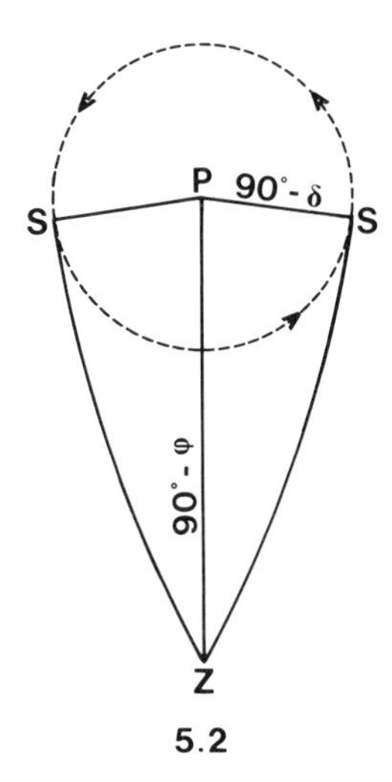

5.2

The spherical geometry of elongation is shown in fig. 5.2, a triangle right-angled at the star. To predict the moment of elongation, the hour–angle P is calculated from $\cos P = \tan \phi \cot \delta$, and the corresponding chronometer reading is calculated. The azimuth is Z or $(360° - Z)$ where $\sin Z = \sec \phi \cos \delta$. At elongation, $\sin h = \sin \phi \operatorname{cosec} \delta$, and the elevation of the star is then greater than the observer's latitude; this may make the method awkward for use in high latitudes.

Since it is impossible to make two settings on a star both at the exact moment of elongation, the observations may be treated by the method in section 5.4.

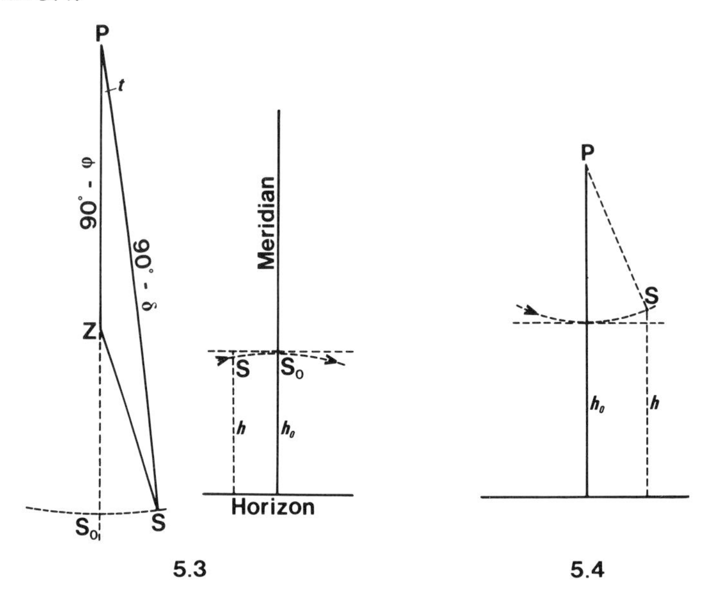

5.3 5.4

5.6 Circum-meridian latitude method

In the Horrebow–Talcott method for finding astronomical latitude the altitude of a star is measured a number of times when it is very close to the point of transit across the meridian. The geometry for the observation of a star at upper transit is indicated in fig. 5.3 and the relevant astronomical triangle has a very small hour–angle, denoted t. It is assumed that the exact time of transit is known and the angles t are calculated from chronometer readings taken at each setting. The object is to use the hour–angles to calculate small additions to the observed altitudes h and convert them to values of h_0, the altitude at the transit point S_0. h_0, δ and ϕ are connected by the simple arithmetical relationship $\phi = 90° + \delta - h_0$.

Since SZ is $90° - h$ we have, by cosine formula:

$$\sin h = \sin\delta \sin\phi + \cos\delta \cos\phi \cos t.$$

For small t, its cosine is closely equal to $1 - \frac{1}{2}t^2$ (t in radian). So we get:

$$\begin{aligned}\sin h &= \sin\delta \sin\phi + \cos\delta \cos\phi - \tfrac{1}{2}t^2 \cos\delta \cos\phi \\ &= \cos(\phi - \delta) - \tfrac{1}{2}t^2 \cos\delta \cos\phi.\end{aligned}$$

Putting $h = h_0 - \mathrm{d}h$, then $\sin h$ is approximated by: $\sin h_0 - \mathrm{d}h \cos h_0$. Also $\phi - \delta = 90° - h_0$, so we find that:

$$\sin h_0 - \mathrm{d}h \cos h_0 = \sin h_0 - \tfrac{1}{2}t^2 \cos\delta \cos\phi$$

whence,
$$dh = \frac{\cos\delta\cos\phi}{2\cos h_0}t^2.$$

This formula is true for dh and t in radian. In practice t is expressed in time–seconds and we want dh in angle–seconds. Putting $dh = dh''/206\,265$ and $t = (15t_s)/206\,265$ and doing the arithmetic gives:

$$dh'' = \frac{\cos\delta\cos\phi}{\cos h_0}\;\frac{15t_s^2}{27502}$$

in which h_0 may be taken as the highest altitude actually observed.

Values of the expression in brackets are tabulated against t_s in the *Star Almanac for Land Surveyors.*

The corrections dh must be calculated separately for each setting on the star.

When the star is at lower transit, fig. 5.4, the corrections must be subtracted from observed h.

CHAPTER 6

Somewhat geometrical

In many ways the geometrical properties of a spherical triangle are similar to those of a plane triangle but there are differences. It is important to note that plane geometry must be regarded as a special or limiting case of spherical geometry. Properties transfer from spherical triangles to plane triangles but not necessarily the other way. An interesting illustration of this is given by the medians. The medians of a spherical triangle are concurrent, as they are in a plane triangle. In a plane triangle each median bisects the area of the triangle but this is not so in a spherical triangle. Of course there is, in a spherical triangle, a line through each vertex to bisect the area and these three lines are concurrent, but they are not the medians.

6.1 Small circle

The intersection of a plane and a sphere is in all cases a circle, and if the plane does not pass through the centre of the sphere the intersection is called a small circle. On a terrestrial globe the parallels of latitude (other than the equator) are small circles, such as indicated by the broken line QQ′ on fig. 6.1. In spherical trigonometry the measure of a small circle is expressed as its angle radius λ. This shows on the surface of the sphere as arcs such as PQ and PQ′ which, being arcs of great circles, can be elements in formulas of spherical trigonometry.

The linear radius of the small circle in its own plane is, of course, $R \sin \lambda$.

6.2 Angle bisectors: In-circle

The angle bisector through vertex A meets the opposite side at angle θ and divides it into parts α_1 and α_2 as in fig. 6.2. Sine formulas for triangles ABL

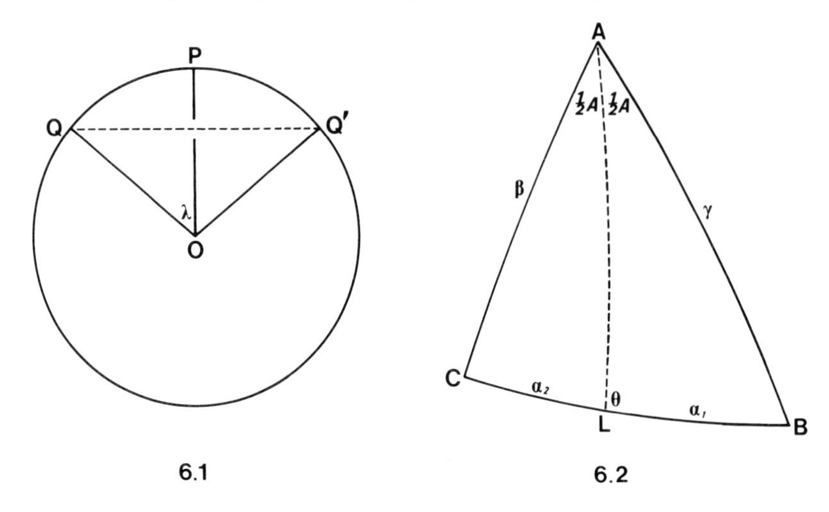

6.1

6.2

and ACL give:

$$\sin\alpha_1 \sin\theta = \sin\gamma \sin\tfrac{1}{2}A$$

and,
$$\sin\alpha_2 \sin\theta = \sin\beta \sin\tfrac{1}{2}A.$$

So by division, $\dfrac{\sin\alpha_1}{\sin\alpha_2} = \dfrac{\sin\gamma}{\sin\beta}$, a formula closely resembling its counterpart for a plane triangle.

As in plane geometry, it is obvious that the three bisectors through the vertices of the spherical triangle must meet at a common point which is equidistant from the three sides (*see* fig. 6.3). There will be a small circle touching the sides at H, J and K. By the tangency property AJ = AK, BK =

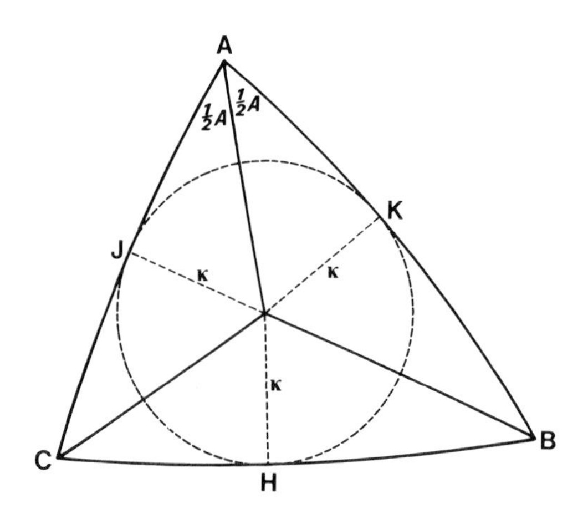

6.3

BH and CH = CJ. Hence (BH + CH + AJ), being the sum of one from each of the pairs of tangents, is equal to $\frac{1}{2}(\alpha + \beta + \gamma) = \sigma$. Then AJ = $(\sigma - \alpha)$ = AK, BK = $(\sigma - \beta)$ = BH and CH = $(\sigma - \gamma)$ = CJ. By right-angled triangle formulas, therefore, the angular radius κ of the in-circle is given by:

$$\begin{aligned} \tan \kappa &= \sin(\sigma - \alpha) \tan \tfrac{1}{2}A \\ &= \sin(\sigma - \beta) \tan \tfrac{1}{2}B \\ &= \sin(\sigma - \gamma) \tan \tfrac{1}{2}C \end{aligned}$$

6.3 Side bisectors: Circum-circle

Again as in plane geometry, the three perpendicular bisectors of the sides of a spherical triangle will obviously meet at a common point which is equidistant from the three vertices. This point is the centre, on the sphere, of the circum-circle, the small circle in which the sphere is intersected by the plane that passes through the vertices. In fig. 6.4, AJ = $\frac{1}{2}\beta$ and AK = $\frac{1}{2}\gamma$, and in the

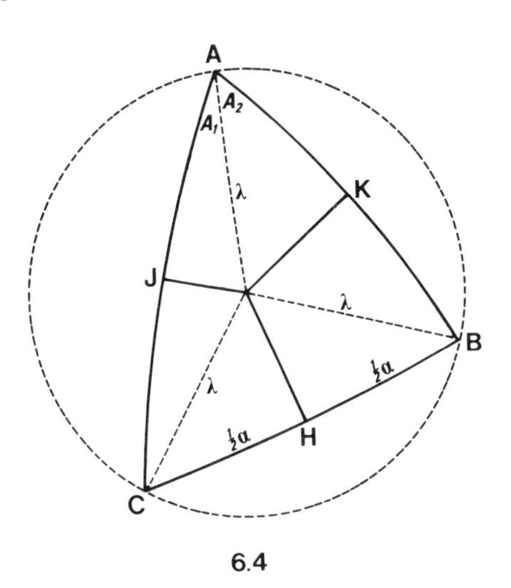

6.4

right-angled triangles containing angles A_1 A_2 we have $\cos A_1 = \tan \frac{1}{2}\beta \cot \lambda$, and $\cos A_2 = \tan \frac{1}{2}\gamma \cot \lambda$. Substituting these into $\cos A = \cos(A_1 + A_2)$ leads to a formula for the spherical circum-radius λ:

$$\tan \lambda = \frac{\sin \alpha}{\sin A} \sec \tfrac{1}{2}\alpha \sec \tfrac{1}{2}\beta \sec \tfrac{1}{2}\gamma.$$

6.4 Transversals concurrent

If lines through the vertices of a spherical triangle are concurrent, opposite pairs of angles at the intersection will be equal, marked ξ, η and ζ in fig. 6.5. The transversals meet the sides at angles θ, ϕ and ψ, and divide them into

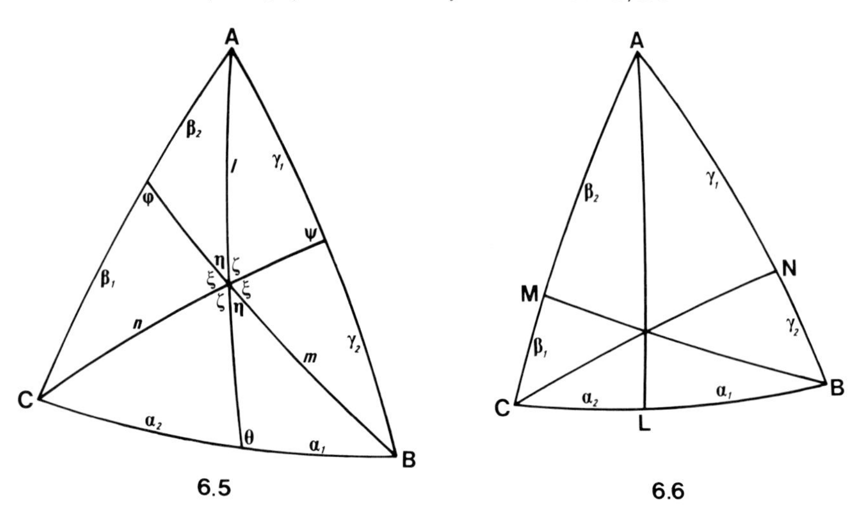

6.5

6.6

portions $\alpha_1, \alpha_2 \ldots$ etc., as shown. Lines from the vertices to the concurrency point are denoted l, m and n. In the triangles containing angle ξ, the sine formulas give: $\sin \beta_1 \sin \phi = \sin n \sin \xi$ and $\sin \gamma_2 \sin \psi = \sin m \sin \xi$. By division and rearrangement:

$$\frac{\sin \beta_1}{\sin \gamma_2} = \frac{\sin n}{\sin m} \times \frac{\sin \psi}{\sin \phi}.$$

Two similar formulas are obtained from the other pairs of triangles, and it is easily seen that if the three are multiplied together, the right-hand sides will cancel out to 1 and the result is:

$$\sin \alpha_1 \sin \beta_1 \sin \gamma_1 = \sin \alpha_2 \sin \beta_2 \sin \gamma_2.$$

This relation is a test for the concurrence of three transversals because, obviously, if either of the transversals is displaced from the concurrence position, the relation will no longer be satisfied.

6.5 Perpendiculars

Lines through the vertices to meet the opposite sides at right angles are shown on fig. 6.6. By the right-angled triangle formulas we have $\sin \alpha_1 = \tan$ AL $\cot B$ and $\sin \alpha_2 = \tan$ AL $\cot C$. So by division, $\sin \alpha_1 / \sin \alpha_2 = \tan C / \tan B$. On multiplying this equation with the two corresponding equations for β_1, β_2, γ_1, γ_2 it is seen that A, B and C cancel out and the condition for concurrence is satisfied.

Also in the triangles ALB and ALC, we have:

$$\sin \text{AL} = \sin \gamma \sin B = \sin \beta \sin C.$$

Hence, $\sin \text{AL} \sin \alpha = \sin \alpha \sin \beta \sin C.$

This can be written as $\sin \alpha \sin \beta \sin \gamma \left(\frac{\sin C}{\sin \gamma}\right)$. Because of the sine formulas, this expression will have the same value for $\sin \text{BM} \sin \beta$ and $\sin \text{CN} \sin \gamma$. Here there is a close similarity with plane triangles in which 'height $\times$ base' is the same for each perpendicular and is equal to twice the area of the triangle. So the formulas $\frac{1}{2} \sin \beta \sin \gamma \sin A$ and $\frac{1}{2} \sin \gamma \sin \alpha \sin B$ and $\frac{1}{2} \sin \alpha \sin \beta \sin C$, may be seen as the spherical equivalents of the plane triangle area formulas $\frac{1}{2}\, b\, c \sin A$, etc.

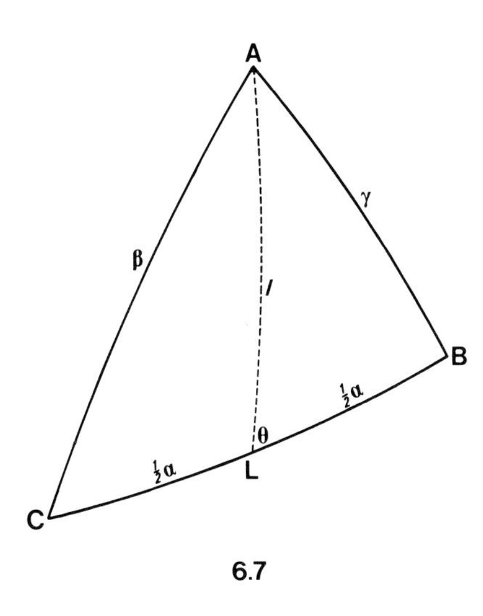

6.7

6.6 Medians

In fig. 6.7, the condition for concurrence is satisfied. The 'lengths' of the medians may be calculated by cosine formulas. Otherwise, the four-part formulas:

$$\cos \tfrac{1}{2}\alpha \cos \theta = \sin \tfrac{1}{2}\alpha \cot l - \sin \theta \cot B$$
$$-\cos \tfrac{1}{2}\alpha \cos \theta = \sin \tfrac{1}{2}\alpha \cot l - \sin \theta \cot C$$

by subtraction lead to:

$$\cot \theta = \tfrac{1}{2}(\cot C - \cot B) \sec \tfrac{1}{2}\alpha$$

then by addition to:

$$\cot l = \tfrac{1}{2} \sin \theta(\cot B + \cot C) \operatorname{cosec} \tfrac{1}{2}\alpha.$$

Though the medians do not bisect the area of the triangle, the two triangles ABL and ACL have the same value, $\frac{1}{2} \sin l \sin \frac{1}{2}\alpha \sin \theta$, for the formula mentioned in the previous section.

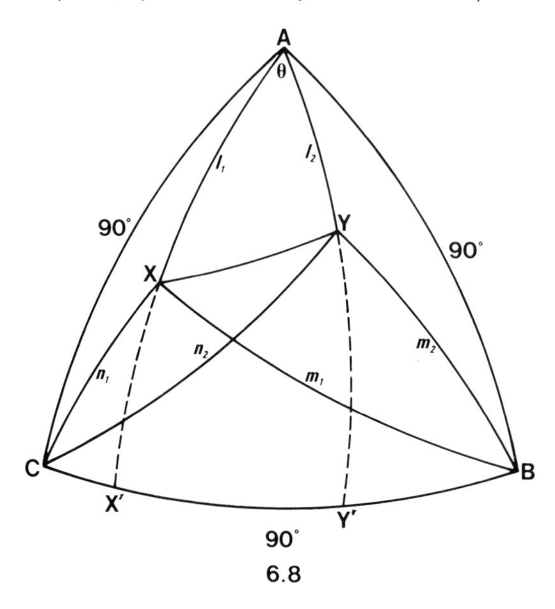

6.8

6.7 Direction cosines

Fig. 6.8 shows the 90° triangle, or octant of the sphere, mentioned in section 3.4. X and Y are any two points in the triangle, and the lines joining them to the vertices of the triangle are denoted as shown. AX and AY are produced to meet BC, so AX′ = AY′ = 90° and X′Y′ = θ.

Now,

$$\begin{aligned}\cos XY &= \cos l_1 \cos l_2 + \sin l_1 \sin l_2 \cos \theta \\ &= \cos l_1 \cos l_2 + \cos XX' \cos YY' \cos \theta.\end{aligned}$$

Noting that triangles BXX′, etc. are right angled, we can write:

$$\cos m_1 = \cos XX' \cos BX'; \cos m_2 = \cos YY' \cos BY'$$
$$\cos n_1 = \cos XX' \cos CX'; \cos n_2 = \cos YY' \cos CY'.$$

But BX′ + CX′ = BY′ + CY′ = 90°, so the formulas for n_1 and n_2 can be written:

$$\cos n_1 = \cos XX' \sin BX'; \cos n_2 = \cos YY' \sin BY'.$$

Hence $\cos m_1 \cos m_2 + \cos n_1 \cos n_2 = \cos XX' \cos YY' (\cos BX' \cos BY' + \sin BX' \sin BY') = \cos XX' \cos YY' \cos (BX' - BY')$

$$= \cos XX' \cos YY' \cos \theta.$$

Thus the expression for cos XY comes to:

$$\cos XY = \cos l_1 \cos l_2 + \cos m_1 \cos m_2 + \cos n_1 \cos n_2.$$

If we regard points A, B and C as representing the directions of three mutually perpendicular axes, then l_1, m_1 and n_1 are the angles which the direction represented by X makes with the three axes. Similarly l_2, m_2 and n_2 are the angles of the direction represented by Y. The above equation gives the cosine of the angle between the two directions X and Y in terms of their individual direction cosines referred to the three rectangular axes.

The equation may be of use in astronomy. For instance, if l_1, m_1 and n_1 are the angles of a direction to a star in a spatial coordinate system, and l_2, m_2 and n_2 are the angles of the direction of the vertical at an observation station, then the 'XY' of the formula is the zenith distance of the star, as mentioned in section 1.6.

When the points X and Y coincide, the formula is simply $\cos^2 l + \cos^2 m + \cos^2 n = 1$, an identity satisfied by any set of direction cosines referred to rectangular axes.

6.8 Side equation

Surveyors are very familiar with networks of points and lines that form the geometrical structure of systems of triangulation and trilateration. Such systems are composed of triangles and polygons of various shapes and sizes that are largely determined by the positions of points on the ground where suitable observations can be made.

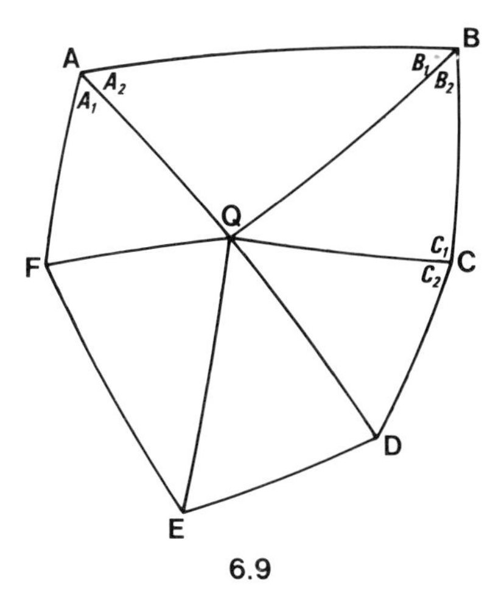

6.9

The side equation is an important relationship between the angles in a polygon on a sphere. A typical piece of triangulation composed of six triangles is illustrated in fig. 6.9. The sine formulas relating to the peripheral angles A_1, A_2, B_1, B_2, C_1, ... are:

$$\frac{\sin A_2}{\sin B_1} = \frac{\sin \mathrm{QB}}{\sin \mathrm{QA}}, \quad \frac{\sin B_2}{\sin C_1} = \frac{\sin \mathrm{QC}}{\sin \mathrm{QB}}, \text{ etc.}$$

It is easily seen that if these equations are multiplied together, putting the product of all the left-hand sides equal to the product of all the right-hand sides, then all the functions sin QA, sin QB, . . . cancel out and the result is:

$$\sin A_1 \sin B_1 \sin C_1 \sin D_1 \sin E_1 \sin F_1 = \sin A_2 \sin B_2 \sin C_2 \sin D_2 \sin E_2 \sin F_2.$$

This is also used in the form:

$$\log \sin A_1 + \log \sin B_1 \ldots + \log \sin F_1$$
$$= \log \sin A_2 + \log \sin B_2 \ldots + \log \sin F_2.$$

A relationship like this holds between the peripheral angles of a polygon of any number of sides.

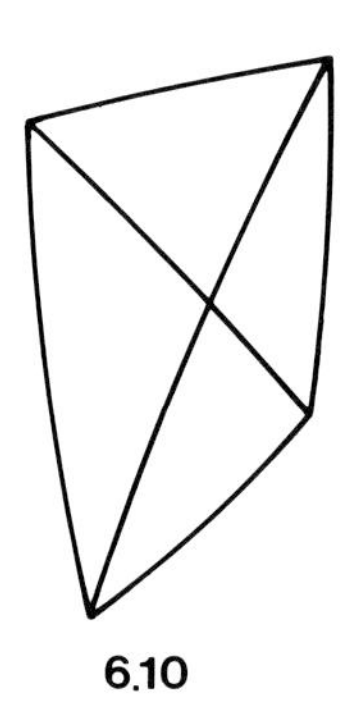

6.10

It is to be noted also that this side equation holds for the eight peripheral angles of a quadrilateral with diagonals, (fig. 6.10), since this is simply a special case of a quadrilateral with a central point. Only the central point is not a station in the triangulation system.

6.9 Spherical excess

It has been stated that the spherical excess of a triangle depends on its area in relation to the area of the whole sphere. In fig. 6.11, ABC is the spherical triangle with its sides extended to the complete great circles as shown. Any great circle through a point passes also through the antipodal point, and it is evident that the three antipodes A′, B′ and C′ form a triangle which is an exact copy of ABC.

The two semicircles ABA′ and ACA′ enclose an area which, by an obvious derivation, is called a *lune*. It is also obvious that the area covered by the lune is a simple fraction of the area of the whole sphere, the fraction being the ratio of angle A to the 360° round A. Reckoning A in radian this fraction is $A/2\pi$. Since the area of the sphere is $4\pi R^2$, the area of the lune is $2AR^2$. Similarly, the area of the lune BAB′CB is $2BR^2$. Add to these the third lune CA′C′B′C, which includes the antipodal triangle, and the total area of the three lunes comes to $2(A + B + C)R^2$.

The three lunes have been ruled so as to distinguish them and it is easily

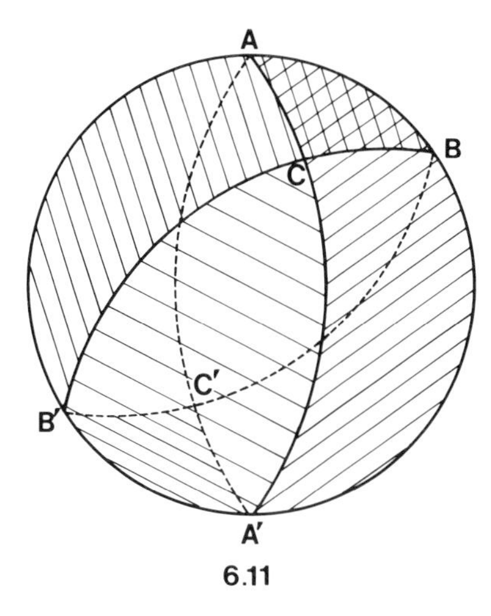

6.11

seen that the total area covered by them is a hemisphere plus two extra triangles. That is:

$$2(A + B + C)R^2 = 2\pi R^2 + 2(\text{triangle area}),$$

which reduces to $(A + B + C - \pi)R^2$ = triangle area. But $(A + B + C - \pi)$ is the spherical excess in radian. Hence, spherical excess = (triangle area)/R^2 in radian. Multiply by 206 265 to convert it to seconds.

On an Earth-size sphere, spherical excess is practically 1″ per 200 square kilometres.

6.10 Legendre's Theorem

We now come to a piece of spherical geometry that has for long been of concern to surveyors and geodesists. The triangles that make up a triangulation figure of the type used as a basis for a mapping system, or for the study of the shape of the Earth, have areas that are quite tiny in comparison with the whole area of the Earth, which is roughly 500 million square kilometres. But a survey triangle covering 1000 km^2 would be described as large, though it has a spherical excess of only about 5 seconds.

Surveyors are concerned with the linear distances between points, and a recurring problem in survey computation is to find the lengths of two sides of a spherical triangle when the three angles and one side are given. Geodesists of long ago, faced by this task, converted the given length to angle measure, using a suitable radius for the sphere, then solved the spherical triangle with standard formulas, and converted the angle measures of the other sides to linear measures. This was evidently a tedious process. Legendre produced another solution which he published in a book on trigonometry early in the nineteenth century.

Imagine a spherical triangle with sides α, β and γ to be made of string, removed from the sphere, and stretched out flat. It will then be a plane triangle with sides of lengths $R\alpha$, $R\beta$ and $R\gamma$ (α, β and γ in radian). The angles of this plane triangle can be calculated by standard formulas of plane trigonometry and they will, of course, add up to exactly 180°. How will they compare with the angles of the original spherical triangle?

Legendre's theorem states that for any triangle appearing in practical surveying or geodesy, the differences between corresponding spherical and plane angles can be taken as equal for all computation purposes. Hence each difference is one-third of the spherical excess.

To find the spherical excess, a few seconds at most, the area may be calculated as if the triangle is plane, or even measured graphically on an accurate scale diagram of the survey system.

So the procedure for the solution of triangles now, is to reduce the angles of the spherical triangle by $\frac{1}{3}$(spherical excess) and solve as for a plane triangle.

In these days of E.D.M., it may be that a surveyor knows the lengths of the three sides and needs to calculate the angles of the spherical triangle. In this case, the angles are calculated precisely by plane formulas, the area is found to sufficient accuracy, and $\frac{1}{3}$(spherical excess) is added to each plane angle.

It is important to note that Legendre's theorem is only a computing device. There is no plane triangle connected in any simple geometrical way with the spherical triangle, having sides of the same lengths.

Though Legendre's theorem does not state that the three differences are exactly equal, the remaining discrepancies are extremely small and quite beyond the limits of accuracy in practical survey measurement, however rigorous. If an enormous spherical triangle, big enough to cover Great Britain, is solved by use of the theorem, the residual errors in the lengths are not more than 2 parts per million! Note also example 2, p. 47.

6.11 Legendre: Proof 1

There are two ways to demonstrate Legendre's theorem. One is to compare the angles of the plane 'string' triangle, mentioned above, with the angles of the spherical triangle. The other is to take a plane triangle with angles $(A - \frac{1}{3}E)$ $(B - \frac{1}{3}E)$ $(C - \frac{1}{3}E)$ (where E is the spherical excess), and work out the proportional relations of its sides and the linear lengths of the sides of the spherical triangle.

If A', B', C' are the angles of the plane 'string' triangle which has sides of lengths proportional to α, β and γ, then by standard plane formula $2\beta\gamma \cos A' = \beta^2 + \gamma^2 - \alpha^2$. By relating this to the spherical cosine formula:

$$\sin \beta \sin \gamma \cos A = \cos \alpha - \cos \beta \cos \gamma,$$

it should be possible to find a formula for the difference $(A - A')$, a difference which, even in the largest triangles likely to be measured in geodetic surveying, will rarely be as much as a couple of seconds.

We are now considering triangles in which the sides α, β and γ in radian measure are very small fractions. On an Earth-size sphere, a side must be almost 64 km long to reckon as 1/100 radian. So, in the spherical formula, series expressions such as $\sin\beta = \beta - \frac{1}{6}\beta^3 \ldots$ and $\cos\beta = 1 - \frac{1}{2}\beta^2 + \frac{1}{24}\beta^4 \ldots$ are substituted, and after some algebraic work stopping at fourth powers, we get:

$$\cos A\,(1 - \tfrac{1}{6}\beta^2 - \tfrac{1}{6}\gamma^2 \ldots)\,\beta\gamma$$
$$= \tfrac{1}{2}\beta^2 + \tfrac{1}{2}\gamma^2 - \tfrac{1}{2}\alpha^2 + \tfrac{1}{24}\alpha^4 - \tfrac{1}{24}\beta^4 - \tfrac{1}{24}\gamma^4 - \tfrac{1}{4}\beta^2\gamma^2 \ldots$$

Dividing by the series in brackets on the left side means multiplying the right side by $(1 + \frac{1}{6}\beta^2 + \frac{1}{6}\gamma^2 \ldots)$ and when this is done the result is:

$$\beta\gamma\cos A = \tfrac{1}{2}\beta^2 + \tfrac{1}{2}\gamma^2 - \tfrac{1}{2}\alpha^2 + \tfrac{1}{24}\alpha^4 + \tfrac{1}{24}\beta^4 + \tfrac{1}{24}\gamma^4 - \tfrac{1}{12}\beta^2\gamma^2 - \tfrac{1}{12}\gamma^2\alpha^2 - \tfrac{1}{12}\alpha^2\beta^2 \ldots$$

Now A' is brought in. The terms $\frac{1}{2}\beta^2 + \frac{1}{2}\gamma^2 - \frac{1}{2}\alpha^2$ on the right are just $\beta\gamma\cos A'$, and further, by squaring this expression, we find that:

$$\beta^2\gamma^2\sin^2 A' = \beta^2\gamma^2 - \beta^2\gamma^2\cos^2 A'$$
$$= -\tfrac{1}{4}\alpha^4 - \tfrac{1}{4}\beta^4 - \tfrac{1}{4}\gamma^4 + \tfrac{1}{2}\beta^2\gamma^2 + \tfrac{1}{2}\gamma^2\alpha^2 + \tfrac{1}{2}\alpha^2\beta^2.$$

Thus the fourth power terms in $\beta\gamma\cos A$ are simply $-\frac{1}{6}\beta^2\gamma^2\sin^2 A'$. On making the substitutions in the equation for $\beta\gamma\cos A$ it is divisible throughout by $\beta\gamma$ and we get:

$$\cos A = \cos A' - \tfrac{1}{6}\beta\gamma\sin^2 A' \ldots$$

Now if we put $A = A' + \theta$, the differential calculus gives $\cos A = \cos A' - \theta\sin A'$, and we see that $\theta = \frac{1}{6}\beta\gamma\sin A'$. But the area of the plane triangle is $\frac{1}{2}R\beta R\gamma\sin A'$ so its spherical excess (section 6.9) is $\frac{1}{2}\beta\gamma\sin A'$. Finally, therefore, $\theta = \frac{1}{3}$(spherical excess).

6.12 Legendre: Proof 2

In the second method of demonstrating Legendre's theorem, let the sides of the triangle be a, b, c. Then a sine formula is:

$$\frac{b}{c} = \frac{\sin(B - \frac{1}{3}E)}{\sin(C - \frac{1}{3}E)}$$

which, since E is very small, can be written:

$$\frac{\sin B - \frac{1}{3}E\cos B}{\sin C - \frac{1}{3}E\cos C}.$$

As noted above, E can be expressed as $\frac{1}{2}\beta\gamma \sin A$, or equally as $\frac{1}{2}\gamma\alpha \sin B$ or $\frac{1}{2}\alpha\beta \sin C$.

Inserting these last two into the sine ratio gives:

$$\frac{\sin B\,(1 - \frac{1}{6}\gamma\alpha \cos B \ldots)}{\sin C\,(1 - \frac{1}{6}\alpha\beta \cos C \ldots)}.$$

In the second order terms we can substitute:

$$\tfrac{1}{12}\gamma^2 + \tfrac{1}{12}\alpha^2 - \tfrac{1}{12}\beta^2 \quad \text{for } \tfrac{1}{6}\gamma\alpha \cos B,$$

and similarly for $\cos C$. Furthermore, in the spherical triangle:

$$\frac{\sin B}{\sin C} = \frac{\sin \beta}{\sin \gamma} = \frac{\beta\,(1 - \frac{1}{6}\beta^2 \ldots)}{\gamma(1 - \frac{1}{6}\gamma^2 \ldots)}.$$

If the substitutions are made and multiplied out, it is found that the series in the numerator and denominator are equal (to second order). In fact $(1 - \frac{1}{12}\alpha^2 - \frac{1}{12}\beta^2 - \frac{1}{12}\gamma^2 \ldots)$, so we are left with $b/c = \beta/\gamma$. That is, the linear lengths of the sides of the spherical triangle are proportional to the sides of a plane triangle that has angles $(A - \frac{1}{3}E)$, $(B - \frac{1}{3}E)$ and $(C - \frac{1}{3}E)$, to second order in the small radian quantities α, β and γ.

Indeed it is evident that the next terms in the series will be of fourth order, indicating that any remaining discrepancies will, in practice, have to be reckoned in thousandths of a second or less.

CHAPTER 7

Small changes

7.1 Errors

A surveyor is for ever having to consider the possible effects of small errors. Any measurement must be assumed to be erroneous, and usually is, and the amount of error is not known. It should be possible, however, for a surveyor to make some estimate of the limits to error which may affect his measurements and then, taking perhaps a rather pessimistic view, to calculate what effects the estimated errors could have on the final results of his work.

Such calculations are useful, not only to give estimates of acccuracy of results, but also to enable the surveyor to use his instruments and arrange his methods so as to minimise the effects of errors, or at least to convince himself that any possible effects are within the specification to which he is working.

7.2 Small changes

An error may be regarded as a small change. If a small change is made in any measurement or other quantity entering into a survey computation, there will be consequential *small* changes in the computed results. A point about small changes is that, for practical purposes, we may assume that the effects are simply proportional to the causes. So we are interested to find a factor, or differential coefficient, by which to multiply the change so as to obtain the effect.

Another thing about small changes is that their effects are additive. For instance if the three sides of a triangle are given, the angles can be computed; then the results of specific small changes in all three sides on one particular angle can be calculated separately and simply added (algebraically) together.

7.3 Variations

In carrying out an adjustment process by a variation method, the effects of small changes have to be calculated, such as the effect of small variations of the coordinates of a point on its distances from surrounding points.

In this chapter we are concerned with calculating the effects of a specific error or change in one measurement when all the other quantities involved are unchanged. Methods of treatment based on statistical rules when, as is usual, all measured quantities are assumed to be subject to random errors, are not dealt with.

7.4 Mathematics or geometry?

In the spherical triangle with sides α, β and γ, suppose that side α is changed by a small amount $d\alpha$ and we require to find the change in angle C. There are two ways of approach to this sort of problem:

(a) A formula for C in terms of α, β, γ is differentiated with respect to C and α, thus obtaining the factor mentioned above.

(b) A diagram is drawn to show the effect geometrically and to suggest a method of calculation.

7.5 Calculus method

For method (a), we require a formula in which α and C are the only variable elements, so that we do not have to consider consequential variations of other elements of the triangle. The relevant formula in the above case is:

$$\cos\gamma = \cos\alpha\cos\beta + \sin\alpha\sin\beta\cos C$$

and the differentials are therefore related by:

$$0 = -\sin\alpha\cos\beta\,d\alpha + \cos\alpha\sin\beta\cos C\,d\alpha - \sin\alpha\sin\beta\sin C\,dC.$$

This gives a connection between the small changes $d\alpha$ and dC, but it is unnecessarily cumbersome. The total coefficient of $d\alpha$ can be reduced (*see* section 4.7) to $-\sin\gamma\cos B$, and after replacing $\sin\beta\sin C$ by $\sin\gamma\sin B$, we find that $dC = -d\alpha\cot B\operatorname{cosec}\alpha$.

The above working is cumbersome because α appears twice in the formula used. It would be simpler if the effect of $d\alpha$ on angle A were asked for, because then the cosine formula for α would be used, and both α and A appear only once in it.

However the working will usually be simpler still by method (b) which is a geometrical treatment of what is essentially a geometrical situation.

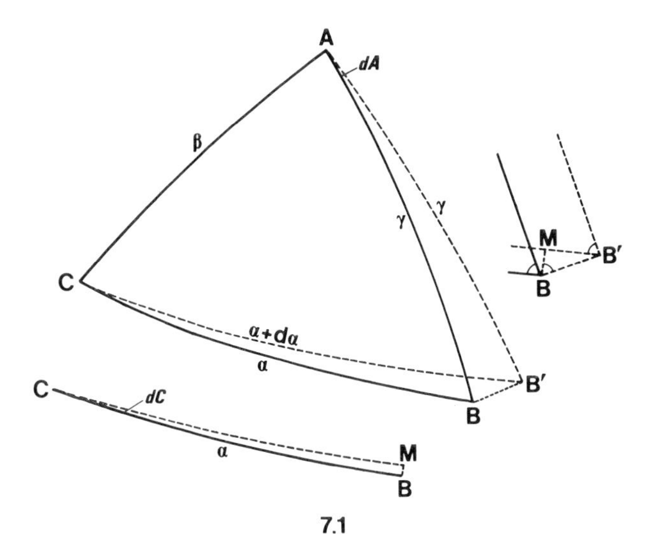

7.1

7.6 Geometrical method

The geometrical approach is illustrated in fig. 7.1. The triangle with sides $(\alpha + \mathrm{d}\alpha)$, β and γ is AB′C. γ is not changed, so AB′ = AB, and BB′ is perpendicular to both. In diagrams like this the small changes have to be enormously exaggerated; changes of the magnitude shown on the diagram could hardly be treated as 'infinitesimal'! BM is perpendicular to CB′ and the infinitesimal right-angled triangle BMB′ can be regarded as plane where only small quantities are involved. The three marked angles are all practically equal to B.

The side MB′ is the change $\mathrm{d}\alpha$, so MB is $\mathrm{d}\alpha \cot B$. Now in the long thin isosceles triangle MCB we apply a sine formula and get $\sin \mathrm{d}C/\sin(\mathrm{d}\alpha \cot B) = \sin 90°/\sin \alpha$. But $\mathrm{d}C$ and $\mathrm{d}\alpha$ are small and practically equal (in radian) to their sines, so $\mathrm{d}C/(\mathrm{d}\alpha \cot B) = 1/\sin \alpha$, that is $\mathrm{d}C = -\mathrm{d}\alpha \cot B \operatorname{cosec} \alpha$, with the minus sign because obviously an increase in α means a decrease in C.

It will be seen that the change in angle A can be calculated similarly, starting with $\mathrm{BB'} = \mathrm{d}\alpha \operatorname{cosec} B$, using the thin triangle ABB′ and leading to $\mathrm{d}A = +\,\mathrm{d}\alpha \operatorname{cosec} B \operatorname{cosec} \gamma$.

In calculating these relations between small changes, an accuracy represented by three or four significant figures should be sufficient since we are dealing with quantities of the order of magnitude of accidental errors, or with variations from initial values that are supposedly very close to the results finally to be obtained.

It will be seen that the geometrical representation of a small error or change, and its effects, requires the construction of a diagram in which the 'original' or 'correct' figure is combined with the 'changed' or 'erroneous' figure in such a way as to show the effects that are required to be calculated.

A suitable juxtaposition must be devised: for instance, if the effect of dα on angle B is required, the 'erroneous' triangle must be placed so that AB is common and there are two sides AC and AC′ of equal lengths β, and sides BC, BC′ enclose the small angle dB.

7.7 An error in astronomy

The examples worked above can be applied directly to astronomical observation using the spherical triangle of section 5.1. With A as the pole, B as observed star and C as the observer's zenith, the dα could be an error in the measured altitude, and then the dC would be the consequential error in the calculated azimuth of the star.

The dA would be the consequential error in the hour–angle hence in the calculated time or longitude. By use of a sine formula we can write dA as dα cosec C cosec β and in astronomical notation (section 5.1) the formula is $\mathrm{d}P = -\mathrm{d}h \operatorname{cosec} Z \sec \phi$, with minus sign because if α increases h decreases. This is the formula which shows that for a given error in the altitude, the consequential error in hour–angle is numerically least when Z is 90° or 270°, that is when the star is on the prime vertical.

7.8 Theodolite errors

Spherical triangles with small angles can be used in studying the effects of small mechanical maladjustments which may exist in a theodolite. Two such maladjustments are (a) transit axis of the telescope not perpendicular to rotation axis of the instrument, (b) line of sight not perpendicular to transit axis.

In making precise observations with a theodolite, the surveyor assumes that both errors exist, but their effects can be cancelled out by the procedure of 'changing face'.

7.9 Transit axis error

The existence of error (a) means that when the instrument is levelled (that is, rotation axis is set truly vertical), the transit axis will not be exactly horizontal. Imagine the instrument to be at the centre of a colossal sphere. In fig. 7.2, V indicates the vertical as well as the rotation axis of the theodolite. H′HH′ represents the plane through the instrument and horizontal there. TT represents the transit axis deviating by a small angle ϵ from perpendicularity to HV. Assuming that error (b) does not exist, the line of sight when the telescope is set at zero elevation will point in the direction indicated by H, but when it is elevated or depressed it will trace out a great circle perpendicular to TT as shown. Let the telescope be set on a point S at elevation h. Then the horizontal circle reading of the theodolite will be

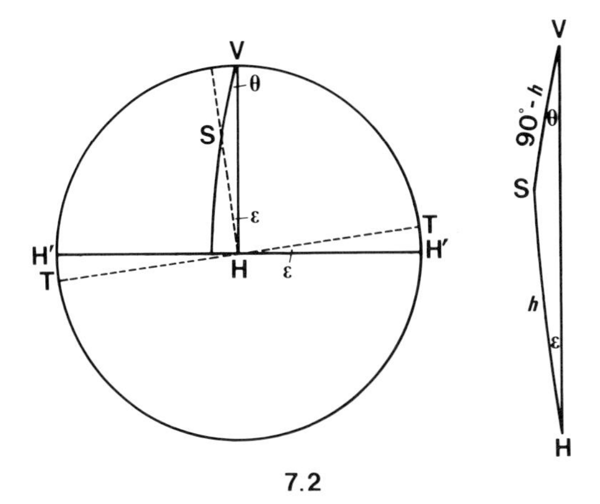

7.2

different from what it would have been if the telescope had transited along HV. The difference is the angle HVS. We thus have a long thin spherical triangle in which the angle at H is ϵ and the angle at V is the effect on the theodolite reading. By sine formula $\sin\theta/\sin\epsilon = \sin \mathrm{HS}/\sin \mathrm{SV}$. Since ϵ and θ are very small, we can take HS $= h$ and SV $= 90° - h$ and write $\theta/\epsilon = \sin h/\cos h$, that is $\theta = \epsilon \tan h$.

The effect of ϵ on a measured angle will be the difference of the values of θ at the two elevations.

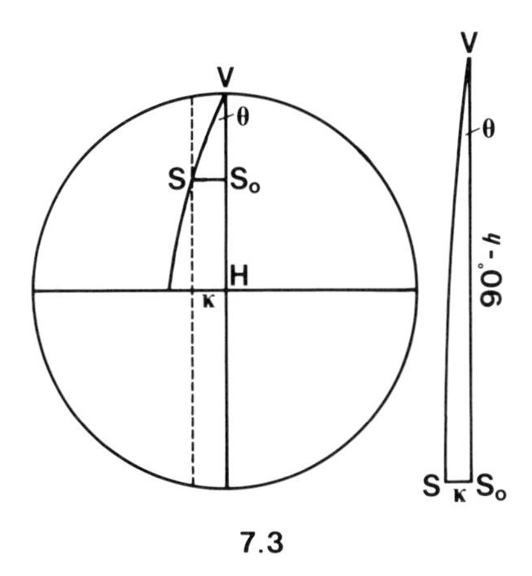

7.3

7.10 Collimation error

The maladjustment (b), horizontal collimation error, is illustrated in fig. 7.3. If there is no transit axis error, the sweep of the line of sight will be a small circle at angle 'distance' κ from the great circle HV. In this case we have the thin triangle shown, and the effect θ by sine formula is $\theta = \kappa \sec h$.

7.11 Dislevelment effects

Another problem involving consideration of small angles occurs when we enquire into the effects of dislevelment of a theodolite, that is when the rotation axis of the instrument is not quite coincident with the gravitational vertical or plumb-line. What effect will this have on measured horizontal angles? (Incidentally, such effects are not eliminated by changing face.)

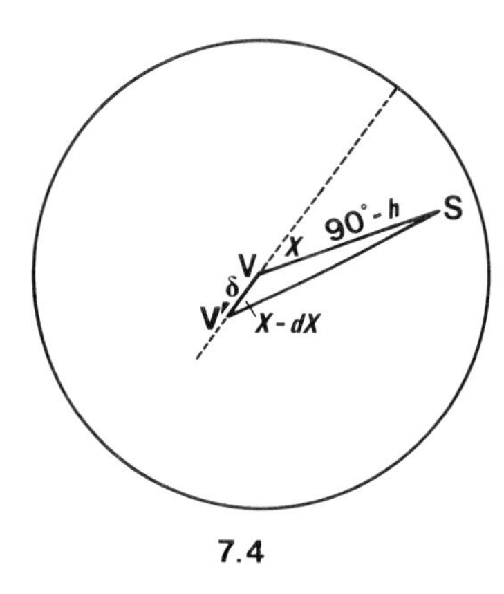

7.4

In fig. 7.4 we are looking down on the theodolite from above. V indicates the true vertical and V′ the direction of the axis of rotation. As usual, the small deviation δ must be enormously exaggerated to show on a diagram. The telescope is set on a signal S at elevation h. Angle X is the horizontal direction of the signal referred to the direction VV′ of the dislevelment. V′S indicates the plane in which the telescope will actually transit, and the angle VV′S is $(X - \mathrm{d}X)$. Then the error in the measured angle between any two signals will be the difference of the errors $\mathrm{d}X$ at the two settings. A four-part formula gives:

$$-\cos\delta\cos X = \sin\delta\tan h - \sin X\cot(X - \mathrm{d}X).$$

We can write $\cos\delta = 1$, $\sin\delta = \delta$ (radian) and by calculus, $\cot(X - \mathrm{d}X) = \cot X + \mathrm{d}X\,\mathrm{cosec}^2 X \ldots$ Putting these in and doing some cancellations gives $\mathrm{d}X = \delta\tan h\sin X$.

7.12 Deviation of the vertical

The foregoing geometry has other applications. Surveying systems such as triangulation are represented as geometrical figures on a spheroidal surface of reference (*see* section 10.1). Through any point on the Earth's surface there will be a line that is perpendicular (normal) to the reference surface, but the direction of this normal will often be slightly different from the direction of the gravity vertical about which a theodolite is levelled. The difference is a deviation of the vertical, seen numerically as the differences between the geodetic and the astronomical latitude and longitude of a point. Horizontal angles are inevitably measured round the true vertical, but for adjusting and computing a geodetic survey we would like to know the angles that would be measured if the theodolite could be set with its axis along the normal. If the deviations are known, the necessary corrections can be calculated by the method of the previous section.

7.13 Black azimuths

Where there is a deviation of the vertical, the astronomical azimuth of a line from one point to another on the Earth will not be quite the same as the geodetic bearing of the corresponding line on the spheroid of reference. But by using an ingenious method proposed by A. N. Black in 1953 (*Empire Survey Review* No. 89), it is possible to obtain a value for the geodetic bearing from astronomical observations. This method uses the fact that the correction term $dX = \delta \tan h \sin X$ is a periodic function of the direction X. Hence, by taking azimuth observations (by the hour–angle method) on a number of stars regularly spaced round the horizon, it is possible to arrange that the average of all the corrections dX is zero, and then it is not necessary to know the deviations.

The geodetic bearing so obtained can then be used as a check in the adjustment and computation of traverses, triangulation, etc.

7.14 Balancing observations

It is often possible to arrange methods of astronomical observation so that certain types of error are self-cancelling or almost so.

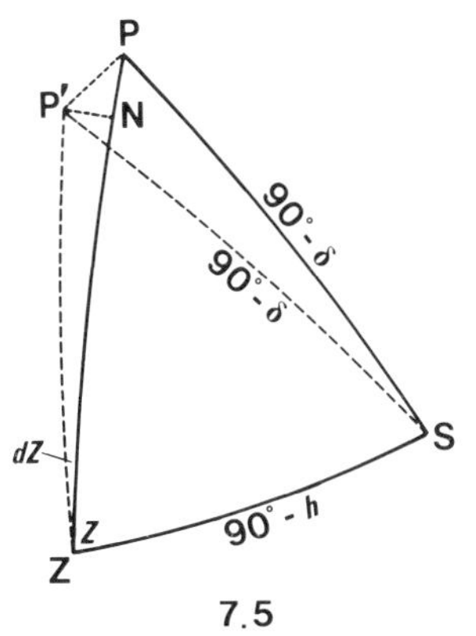

7.5

For instance in finding the azimuth by altitude (section 5.3), the astronomical latitude should be used in the calculation, but a latitude taken from a map may be different on account of a deviation of the vertical. Fig. 7.5

shows the 'correct' and 'erroneous' triangles for this case. ZP is the true co-latitude but ZP′S is the calculated triangle. PN is $d\phi$, the deviation in latitude, and by a working similar to that of section 7.6, it is found that $dZ = d\phi \cot P \sec \phi$. Thus the star's azimuth actually calculated will be $Z + dZ$.

Now if a star could be observed west of the meridian in the position symmetrically opposite to S, the same angle $Z + dZ$ would be calculated but the azimuth of the star would then be $360° - Z - dZ$, and so the mean of the azimuths would be free of the error.

Another error that would cancel out is a difference between calculated actual and refraction.

Examples 1

1. Calculate the angles of the spherical triangle that has sides 30°, 40° and 50°. (Use inverse cosine formulas and check sine formulas.)
2. Calculate the angles of the spherical triangle that has sides 3°, 5° and 7°, and compare with the angles of a plane triangle having sides 3, 5 and 7 units.
3. Calculate the sides and angles of the spherical triangle formed by joining the middle points of the sides of the triangle of question 1. (Use cosine formulas to get the sides.)
4. A spherical equilateral triangle has sides α. Show that the angles A are given by $\cos A = \tan \frac{1}{2}\alpha \cot \alpha$. Find formulas for $\frac{1}{2}A$ and for the circum-radius of the triangle. (In this case, lines from the vertices through the central point of the triangle obviously meet the opposite sides at right angles. Right-angled triangle formulas can be used.)
5. Calculate the sides of the spherical triangle having angles 30°, 45° and 120°. (Use inverted polar cosine formulas and check sine formulas.)
6. Solve the spherical triangle in which $\alpha = 45°$, $\beta = 60°$ and $A = 30°$, (an 'ambiguous' case with two solutions).
7. Solve the spherical triangle in which $A = 30°$, $B = 45°$, $\alpha = 30°$. (Since $\alpha = A$ this is an example of a triangle in which the sine formulas are simply $1 = 1 = 1$. Hence each side must be equal or supplementary to the opposite angle. The third angle must be greater than 90°, so take $C = 180° - \gamma$ and write the cosine formula for side γ.) Show also that $\tan \gamma = (\sin \alpha + \sin \beta)/(\cos \alpha \cos \beta)$.
8. Solve the spherical triangle in which $\alpha = 60°$, $B = 45°$, $C = 135°$. (In plane trigonometry this construction would produce two parallel lines through B and C. On a sphere any two great circles must intersect at two antipodal points, just as two meridians intersect at the poles. From a

four-part formula show that $\cot \beta = \cos B \tan \frac{1}{2}\alpha$. Since B and C are supplementary, β and γ are also supplementary.) Prove that the median from A to the middle point of BC is 90°.

9. A surveyor at latitude 8° 51′ 08″ north observes a star east of his meridian and finds the true altitude to be 39° 02′ 57″. The declination of the star was 9° 40′ 41″ south. Calculate the hour–angle and the azimuth of the star.
10. An observer is at latitude 52° 12′ 06″ north. What will be the altitude and hour–angle of a star of declination 23° 54′ 10″ north when it is on the prime vertical (i.e. azimuth 90° or 270°)?
11. For an observer at latitude 26° 44′ 42″ north, what will be the hour–angle, altitude and azimuth of a star of declination 61° 53′ 45″ north when it is at elongation east of the meridian?
12. A surveyor at latitude 52° 12′ 06″ north measures the horizontal angle between a star and a ground signal and finds it to be 31° 24′ 35″ reckoned clockwise from signal to star. At the same time he recorded the precise chronometer readings and calculated that the hour–angle of the star was 59° 17′ 38″ west of the meridian. The declination of the star was 27° 11′ 18″ north. What was the azimuth of the ground mark?
13. A spherical triangle is solved for given values of A, β and γ. Show that if a small change dA is made in angle A, the effect on side α is $dA \sin p$, where p is the perpendicular from A to the side BC. Find formulas for the changes in angles B and C. What would be the error in the calculated azimuth in question 12, due to an error of 15″ in the hour–angle?

Part II Spheroid

CHAPTER 8

Ellipse geometry

A spheroid is the shape formed by rotating an ellipse about its minor axis. The reason why this form of surface is of interest to surveyors and geodesists is of course that the Earth's overall shape – figure of the Earth – can be approximated very closely to that of a spheroid. But we have to consider spheroids that differ only very slightly from the true spherical shape – by about $\frac{1}{3}$ of 1%, in fact.

8.1 Mapping

We regard a map as a representation of the relative horizontal positions of features on the Earth's surface. Survey measurements are made at points on the Earth's physical surface at various heights above sea level. Mean sea level is a natural reference datum for heights but its shape has local irregularities determined by the geological structure of the Earth's crust. However, with the aid of some astronomical observations it is possible to work out the dimensions and position of a spheroid so that it is a close fit to the mean sea level (MSL) over a limited area of the Earth, such as the extent covered by a single precise mapping system.

In reality therefore the mapping situation is as indicated in fig. 8.1. The positions of points A, B, . . . on the Earth's surface are represented by points A_0, B_0, . . . on the adopted spheroidal reference surface (SPH). Precise mathematical computations based on the geometry of a spheroid can then be applied. In turn, the geometry on the reference surface can be accurately transformed to a plane surface by means of a map projection, as described in Part III.

On a diagram like fig. 8.1. the irregular 'undulations' of the mean sea level surface have to be enormously exaggerated; nowhere on Earth is the mean

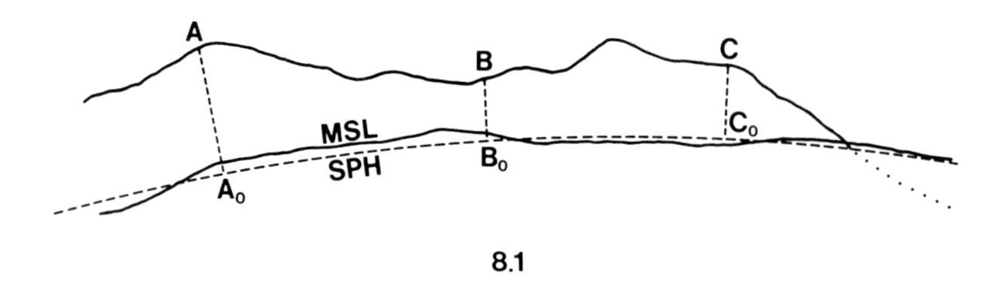

8.1

sea level concave upwards, or anywhere near that condition!

8.2 Spheroid geometry

The geodetic surveyor must therefore take account of the geometry of a spheroidal shape. Any spheroid that is a reasonably good fit to the Earth is so nearly spherical that a scale model as wide as this book would not visibly differ from a sphere. Moreover, the triangles that make up a survey system are relatively so small that some of the relationships of spherical trigonometry can be applied to them with quite sufficient accuracy.

However, a surveyor is concerned with lengths as well as angles, and in computations on a spheroid he has to use formulas that contain linear quantities.

Spheroid geometry is obviously to be derived from the properties of an ellipse. In past times, much use had to be made of the fact that the ellipse had very small eccentricity, but with modern computing facilities the calculation of geodetic quantities to the necessary precision poses fewer problems.

8.3 Ellipse

Of many ways for generating an elliptic shape, perhaps the one most relevant in the present context is a one-way reduction of a circle – a kind of foreshortening.

In diagrams illustrating spheroid geometry the ellipticity must be highly exaggerated. In fig. 8.2, FOE is a diameter of a circle of radius a. Chords like A′D′ perpendicular to FE are reduced in the constant ratio b/a, that is $KA/KA' = KD/KD' = b/a$. This produces an ellipse. In particular $OP = b$ and we describe the ellipse as having semi-axes of lengths a and b.

A measure of the departure of an ellipse from circularity is the proportional reduction of a to b. The fraction $(a - b)/a$ or $1 - (b/a)$ is called the *ellipticity* or *flattening*, usually denoted by f.

In a rectangular coordinate system with OE and OP as axes, the coordinates of A are $x = OK = a \cos\theta$ and $y = KA = (b/a)\, KA' = (b/a)\, a \sin\theta = b \sin\theta$. Thus $(x/a)^2 + (y/b)^2 = \cos^2\theta + \sin^2\theta = 1$, which is the familiar cartesian formula for an ellipse.

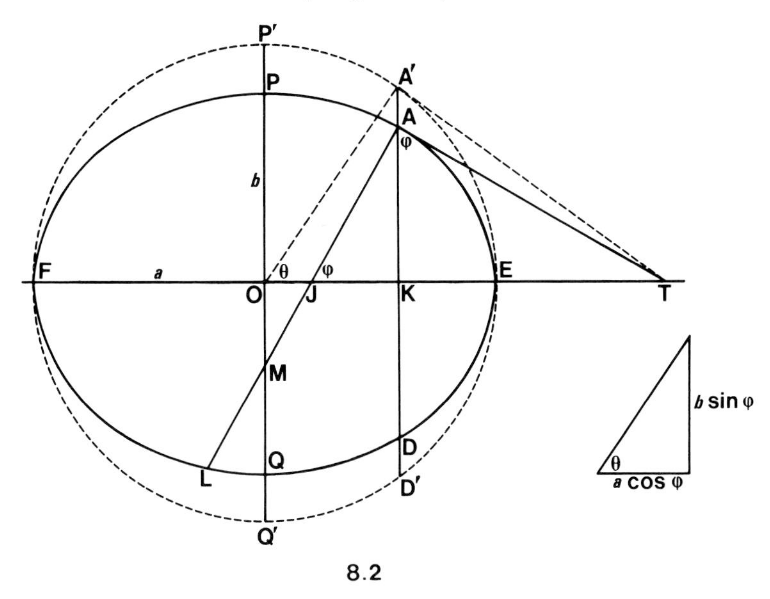

8.2

8.4 Geodetic coordinates

The line A′T is the tangent of the circle at A′. The scale reduction of the circle, to produce the ellipse, does not affect the property of tangency, so AT is the tangent of the ellipse at A. This is evident also if one considers the ellipse and its tangent as the visual effect of rotating the circle and its tangent about the line FET. Of course the radius OA′, on rotation, will no longer appear to be perpendicular to the tangent. The ellipse normal perpendicular to AT is shown as AML.

Angle ϕ is called the *geodetic* (or *spheroidal*) *latitude* of point A. It is to be noted that this angle is an indicator of the position of A on the ellipse and is not to be confused with astronomical latitude.

When the ellipse is rotated about PQ to generate the spheroid surface the angles round this axis between different positions of the elliptical meridians are differences of longitude, just as on a sphere. Thus there can be established a coordinate system of geodetic latitude and longitude indicating positions on the spheroid surface.

In geodetic working it is convenient to express formulas of elliptic geometry in terms of ϕ rather than using an angle at the centre of the ellipse. The angle θ will be seen in some writings on geodetic geometry and is called the *reduced latitude*. The angle AOE (not shown on the diagram) is called *geocentric latitude*.

8.5 Normal. Eccentricity

The length of AM on the normal from A to the minor axis is an important quantity in geodetic computation (*see* section 9.8). It is usually denoted by ν. Referring to the diagram, we see that OK $= a\cos\theta = \nu\cos\phi$, so to express ν in terms of ϕ we need a relation between ϕ and θ. This is obtainable from the triangles KAT and KA'T in which the angles at A and A' are ϕ and θ respectively. KA $=$ KT $\cos\phi$ and KA' $=$ KT $\cot\theta$, but by construction KA/KA' $= b/a$ and we get the relationship $\cot\phi/\cot\theta = b/a$ or $\tan\theta = (b/a)\tan\phi$. This relation is illustrated on fig. 8.2 by the separate triangle which has its hypotenuse of length:

$$(a^2\cos^2\phi + b^2\sin^2\phi)^{\frac{1}{2}} = a[\cos^2\phi + (b^2/a^2)\sin^2\phi]^{\frac{1}{2}}$$
$$= a\left\{1 - [1 - (b^2/a^2)]\sin^2\phi\right\}^{\frac{1}{2}}.$$

$[1 - (b^2/a^2)]$ is e^2 where e is the eccentricity of the ellipse. Thus we see from the triangle that $\cos\theta = \cos\phi\,(1 - e^2\sin^2\phi)^{-\frac{1}{2}}$ and hence

$$\nu = a(1 - e^2\sin^2\phi)^{-\frac{1}{2}}.$$

As ϕ goes from 0° to 90°, ν increases from a to a^2/b. To make our formulas look neater we can write a/ν instead of the clumsy $(1 - e^2\sin^2\phi)^{\frac{1}{2}}$.

8.6 Coordinates

To express the rectangular coordinates of A in terms of ϕ, we already have OK $= x = \nu\cos\phi$. Multiplying the ellipse equation by b^2 gives:

$$(b^2/a^2)x^2 + y^2 = b^2 = a^2(1 - e^2).$$

Then,

$$(1 - e^2)x^2 + y^2 = (1 - e^2)\nu^2(1 - e^2\sin^2\phi).$$

Substitute for x^2 and after some simplification we get:

$$y = \nu(1 - e^2)\sin\phi.$$

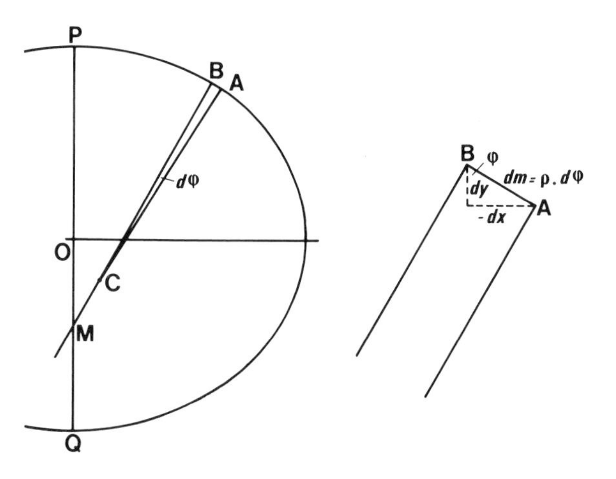

8.3

8.7 Radius of curvature

Another important quantity associated with a point on the ellipse is the length of the radius of curvature, usually denoted by ρ. This is the radius of the circle of 'best fit' to the ellipse in the neighbourhood of the point; the centre of the circle is of course on the ellipse normal. We can think of the normals at two points A and B that are very close together, intersecting at the point C on fig. 8.3. The angle $\mathrm{d}\phi$ between the two normals is the difference of the latitudes of A and B, and the radius ρ (= CA) connects $\mathrm{d}\phi$ with the linear distance AB (= $\mathrm{d}m$) through the relation $\mathrm{d}m = \rho\,\mathrm{d}\phi$ ($\mathrm{d}\phi$ in radian).

A formula for ρ in terms of ϕ is obtainable from the ellipse equation by standard calculus methods, and in other ways. For instance, in the detail on fig. 8.3, we see that $\mathrm{d}m \cos\phi = \mathrm{d}y$, hence $\rho = (\mathrm{d}m/\mathrm{d}\phi) = (\mathrm{d}y/\mathrm{d}\phi)\sec\phi$. By differentiating the formula for ν we find that:

$$\begin{aligned}\mathrm{d}\nu/\mathrm{d}\phi &= ae^2 \sin\phi\cos\phi(1 - e^2\sin^2\phi)^{-\frac{3}{2}}\\ &= \nu e^2 \sin\phi\cos\phi(1 - e^2\sin^2\phi)^{-1}.\end{aligned}$$

But

$$\begin{aligned}y &= \nu(1 - e^2)\sin\phi, \text{ so:}\\ \mathrm{d}y/\mathrm{d}\phi &= \nu(1 - e^2)\cos\phi + \nu e^2(1 - e^2)\sin^2\phi\cos\phi(1 - e^2\sin^2\phi)^{-1}\\ &= \nu(1 - e^2)\cos\phi(1 - e^2\sin^2\phi)^{-1}(1 - e^2\sin^2\phi + e^2\sin^2\phi)\\ &= \nu(1 - e^2)\cos\phi(1 - e^2\sin^2\phi)^{-1}.\end{aligned}$$

Hence

$$\begin{aligned}\rho = (\mathrm{d}y/\mathrm{d}\phi)\sec\phi &= \nu(1 - e^2)(1 - e^2\sin^2\phi)^{-1}\\ &= a(1 - e^2)(1 - e^2\sin^2\phi)^{-\frac{3}{2}}.\end{aligned}$$

8.8 Series formulas

The lengths ρ and ν are of the same order of magnitude as the semi-axes of the ellipse, and this means in practice about six million metres. The formulas given above can be written as an infinite series by use of the Binomial theorem. For instance:

$$\nu = a + \tfrac{1}{2}ae^2\sin^2\phi + \tfrac{3}{8}ae^4\sin^4\phi + \tfrac{5}{16}ae^6\sin^6\phi \ldots$$

In the days when a computer's best friend was his set of 7-figure logarithmic tables, the series formulas had to be used if a precision closer than one metre was required. For any practical spheroid ae^2 is about 40000 metres, so the term in $\sin^2\phi$ can be calculated to centimetre accuracy or better with 7-figure tables. Successive terms in the series rapidly become smaller; ae^4 is about 300 m and ae^6 is about 2 m; the term in $\sin^8\phi$ is, in practice, negligible since its full numerical coefficient is only about 3 millimetres.

The expansion for ρ is:

$$(a - ae^2)(1 + \tfrac{3}{2}e^2\sin^2\phi + \tfrac{15}{8}e^4\sin^4\phi + \tfrac{35}{16}e^6\sin^6\phi + \ldots).$$

Direct computation of ρ and ν from the closed formulas to centimetre accuracy requires 9- or 10-figure capacity, which presents no problem nowadays, being possible on many pocket-size calculators.

8.9 Normal chord

A few more details of the geometry of an ellipse should be mentioned.

Since KA (fig. 8.2) is $\nu(1 - e^2)\sin\phi$, it follows that AJ is $\nu(1 - e^2)$ and JM is νe^2. Hence all normals are divided in the ratio $(1 - e^2)/e^2$ at the points where they cross the major axis OE. Further, OJ is $\nu e^2 \cos\phi$ so the perpendicular distance from O to the normal is $\nu e^2 \sin\phi \cos\phi$. In an Earth-fitting spheroid this has a maximum of about 21 km.

A formula for the full length of the normal chord, AL, can be found as follows. Let the length be c; then the rectangular coordinates of L are $(\nu\cos\phi - c\cos\phi)$ and $(\nu(1 - e^2)\sin\phi - c\sin\phi)$. These must satisfy the ellipse formula, and if they are substituted in the equation $x^2(1 - e^2) + y^2 = \nu^2(1 - e^2)(1 - e^2\sin^2\phi)$, the terms not containing c cancel out:

$$c = 2\nu(1 - e^2)(1 - e^2\cos^2\phi)^{-1}.$$

8.10 Arc length

Another quantity for which numerical values are needed, usually in connection with computations of projection coordinates, is the length of the ellipse curve from equator E to point A at geodetic latitude ϕ. This cannot be expressed as a finite formula. From the relation $dm = \rho\, d\phi$, the curve length EA is found by integration:

$$m = a(1 - e^2)\int_0^{\phi}(1 + \tfrac{3}{2}e^2\sin^2\phi + \ldots)\, d\phi.$$

To integrate this series, the powers of $\sin^2\phi$ are replaced by cosines of multiples of ϕ using the identities:

$$\begin{aligned}
2\sin^2\phi &= 1 - \cos 2\phi \\
8\sin^4\phi &= 3 - 4\cos 2\phi + \cos 4\phi \\
32\sin^6\phi &= 10 - 15\cos 2\phi + 6\cos 4\phi - \cos 6\phi, \text{ etc.}
\end{aligned}$$

After substitution and rearrangement the terms are integrable and the result is:

$$\begin{aligned}
m = (b^2/a)[&(1 + \tfrac{3}{4}e^2 + \tfrac{45}{64}e^4 + \tfrac{175}{256}e^6 \ldots)\,\phi \\
&- (\tfrac{3}{8}e^2 + \tfrac{15}{32}e^4 + \tfrac{525}{1024}e^6 \ldots)\sin 2\phi \\
&+ (\tfrac{15}{256}e^4 + \tfrac{105}{1024}e^6 \ldots)\sin 4\phi \\
&- (\tfrac{35}{3072}e^6 \ldots)\sin 6\phi \ldots]
\end{aligned}$$

or:

$$\begin{aligned}
m = a[&(1 - \tfrac{1}{4}e^2 - \tfrac{3}{64}e^4 - \tfrac{5}{256}e^6 \ldots)\,\phi \\
&- (\tfrac{3}{8}e^2 + \tfrac{3}{32}e^4 + \tfrac{45}{1024}e^6 \ldots)\sin 2\phi \\
&+ (\tfrac{15}{256}e^4 + \tfrac{45}{1024}e^6 \ldots)\sin 4\phi \\
&- (\tfrac{35}{3072}e^6 \ldots)\sin 6\phi \ldots]
\end{aligned}$$

8.11 Finite arc

The equation $\mathrm{d}m = \rho\ \mathrm{d}\phi$ expresses the relation between infinitesimal differences of latitude and curve-length. It is of some practical interest to ask what sort of inaccuracy will result if the infinitesimals are treated as small finite differences, Δm and $\Delta\phi$, and some mean value is taken for ρ.

When ρ as the radius of curvature at latitude ϕ, let us consider the length of the ellipse between latitudes $(\phi + \frac{1}{2}\delta)$ and $(\phi - \frac{1}{2}\delta)$. Writing the formula for m concisely as:

$$m = P\phi - Q \sin 2\phi + R \sin 4\phi \ldots$$

The values of m at $(\phi + \frac{1}{2}\delta)$ and $(\phi - \frac{1}{2}\delta)$ are:

$$P(\phi + \tfrac{1}{2}\delta) - Q \sin (2\phi + \delta) + R \sin (4\phi + 2\delta) \ldots$$
$$P(\phi - \tfrac{1}{2}\delta) - Q \sin (2\phi - \delta) + R \sin (4\phi - 2\delta) \ldots$$

The difference of these, after applying some standard trigonometrical transformation, is:

$$\Delta m = P\delta - 2Q \cos 2\phi \sin \delta + 2R \cos 4\phi \sin 2\delta \ldots$$

But $\rho = \mathrm{d}m/\mathrm{d}\phi$, so by differentiating the series for m:

$$\rho = P - 2Q \cos 2\phi + 4R \cos 4\phi \ldots$$

If the arc of ellipse were taken as a portion of a circle of radius ρ, its length would be $\rho\delta$, and the error in making the assumption would then be $(\rho\delta - \Delta m)$, which from the above formulas is seen to be:

$$- 2Q \cos 2\phi(\delta - \sin \delta) + 2R \cos 4\phi\ (2\delta - \sin 2\delta) \ldots$$

Now $(\delta - \sin \delta)$ is $(\frac{1}{6}\delta^3$ − smaller terms) and $2Q$ is $(\frac{3}{4}ae^2$ + smaller terms) so the coefficient of $\cos 2\phi$ above is $(\frac{1}{8}ae^2\delta^3$ − smaller terms).

Take δ as $\frac{1}{60}$ radian, somewhat smaller than 1° and equivalent to a linear distance of about 106 km. Then the coefficient is about 25 millimetres. In middle latitudes the factor $\cos 2\phi$ will reduce this considerably. However, since the error is proportional to δ^3 it is evident that the simple formula $\rho\delta$ should not be used beyond about 1° if precision within a few centimetres is to be assured.

8.12 Polar equation

Geodesists making measurements with the aid of artificial satellites are interested in the direct straight-line distances between points on the Earth, and the distances involved in this kind of geodesy are very much greater than those that make up the geometry of a land-based survey system. In working with chord lines between points on a spheroid, it may be useful to express an ellipse in a polar type of equation.

This is illustrated in fig. 8.4. AT is the tangent at the fixed point A at

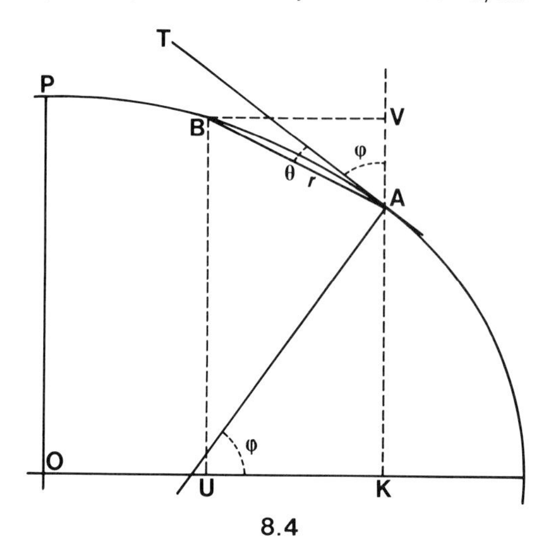

8.4

spheroidal latitude ϕ. Another point B is at chord distance r from A and the chord makes angle θ with AT. Thus r and θ are polar coordinates of B. The rectangular coordinates of B are OU and UB, and we see from the diagram that:

$$OU = OK - KU = \nu \cos\phi - r \sin(\phi + \theta)$$

and

$$UB = KA + AV = \nu(1 - e^2)\sin\phi + r\cos(\phi + \theta).$$

Substituting these into the ellipse equation:

$$(1 - e^2)x^2 + y^2 = \nu^2(1 - e^2)(1 - e^2\sin^2\phi)$$

leads, after some cancellation and simplification, to the polar equation:

$$r(1 - e^2\sin^2(\phi + \theta)) = 2(1 - e^2)\,\nu\sin\theta.$$

Putting $\theta = 90°$ in this equation gives again the formula for the chord length AL (section 8.9).

8.13 Figure of the Earth

Under the law of gravitation a rotating Earth must 'bulge' round the equator. Isaac Newton estimated a flattening of $\frac{1}{230}$. Early in the eighteenth century, the French Academy of Sciences sent geodetic survey expeditions to Peru and Lapland to measure the radius of curvature of the meridian at two widely differing latitudes and they came up with a flattening of $\frac{1}{310}$. In the nineteenth century, many countries of the world were busy making accurate maps and for this purpose each national survey organisation worked out a 'spheroid of best fit' for its area of operations. Hence the numerous different specifications (of a and b, or a and f) to be found in the early literature of geodesy.

In 1924 there was a proposal for adoption of an 'international spheroid' with $a = 6\,378\,388$ metres and $f = \frac{1}{297}$. This was in fact the specification derived by Hayford for a fit to North America. Subsequent investigations have shown that this value for a is too big. Anyway, national mapping organisations continue, with good reason, to use spheroids suitable to their own regions.

Nowadays, however, geodesists are able to measure the Earth–as–a–whole with the aid of artificial satellites, either by the use of geometrical methods similar to triangulation and trilateration, or by observing the effects on satellite orbits due to the shape of the pattern of gravitational forces in the space around the Earth.

One global spheroid that has come from the results of observations of satellites is $a = 6\,378\,160$ m and $f = 1/298.25$, giving $b = 6\,356\,775$ m and $e^2 = 0.006\,694\,542$. The radius ρ is $6\,335\,461$ m at the spheroid equator and $6\,399\,617$ m at the poles. In this specification the formula for arc length m is:

$$6\,367\,471.849\,\phi - 16\,038.955 \sin 2\phi + 16.833 \sin 4\phi - 0.022 \sin 6\phi \text{ metres.}$$

It is interesting to note that if ϕ is put at 90° $(=\frac{1}{2}\pi)$ all the trigonometrical terms vanish and we find that the equator–pole distance is $10\,002\,001.4$ m; a ten-millionth of the semi-meridian was the primitive definition of the metre.

CHAPTER 9

Spheroid geometry

In this chapter and the next, some account is given of the geometry and formulas relevant to the surveyor's treatment of systems of triangulation, trilateration and traverse represented on a spheroidal surface. These systems are made up of lines that are of necessity short enough to be 'intervisible' on the ground. There are two kinds of problem to be considered: the 'internal' relationships of angles and distances in spheroidal geometrical figures; and the calculation of spheroidal latitudes and longitudes of points so as to place the figures in position on the adopted spheroid of reference.

9.1 Line of sight

In real life, a surveyor makes measurements on the irregular surface of the Earth, sometimes at considerable heights above the reference surface, and there are problems in relating such measurements correctly to the corresponding geometry on the adopted spheroid. However, what we now have to think about is the sort of geometry that would be relevant to the work of an imaginary surveyor on a perfectly spheroidal Earth. What, for instance, do we mean by 'the distance between two points' on a spheroid and what is a 'spheroidal triangle'?

On an ideal spheroidal Earth, the direction of gravity, that is the vertical, would be everywhere perpendicular to the physical surface. The procedure of 'levelling' a theodolite means adjusting its position so that its rotation axis is vertical, and thus would coincide with the geometrical normal to the surface, the line AM in fig. 9.1.

Suppose that the telescope is directed to another point B and the horizontal clamps are closed; then if the telescope is rotated on the transit axis, the line of sight will sweep out a plane; the plane defined by the point B and

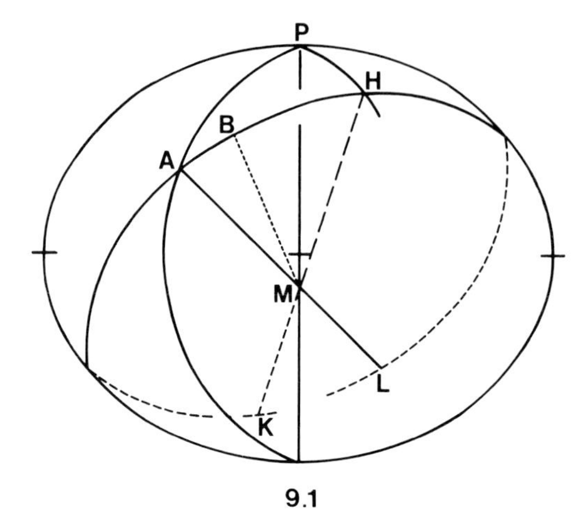

9.1

the normal AM. Any plane section of a spheroid is an ellipse (or in special positions, a circle). The whole section by the plane BAM is indicated on the diagram. Points B and M are joined, but it is to be noted that BM is not the normal at B.

If the surveyor at A 'sets out' the line from A to B by a series of markers sighted through the telescope, he will set out the elliptical curve as shown.

Of course, no surveyor would be able to sight the enormous distance indicated by AB on the diagram, but we are here really considering a problem of spheroid geometry rather than practical surveying! In practice, except at very short distances, there would have to be a signal above the surface at B. The effect of 'signal height' on the line set out from A can be calculated as shown in section 10.11.

9.2 Normal plane sections

Thus we see that the surveyor on a spheroid measuring horizontal angles is concerned with sections of the spheroid surface where it is cut by planes containing the normal at the theodolite station. If he reads the horizontal circle, unclamps and sets on another point C, the difference of the readings is a measure of the angle between the planes BAM and CAM. One vertical plane at A is the meridian itself, and the angle between the meridian plane and the plane BAM is the (spheroidal) bearing of B at A. Another vertical plane is the one perpendicular to the meridian, bearing 90° and 270°, sometimes referred to as the prime vertical section. This is obviously symmetrical about the chord AL which is therefore the minor axis of the ellipse of the section.

We now suppose that the theodolite is set up at B and directed at A. The relevant plane section is now the plane defined by the point A and the normal BN; so the 'set out' curve from B to A will not be quite the same as

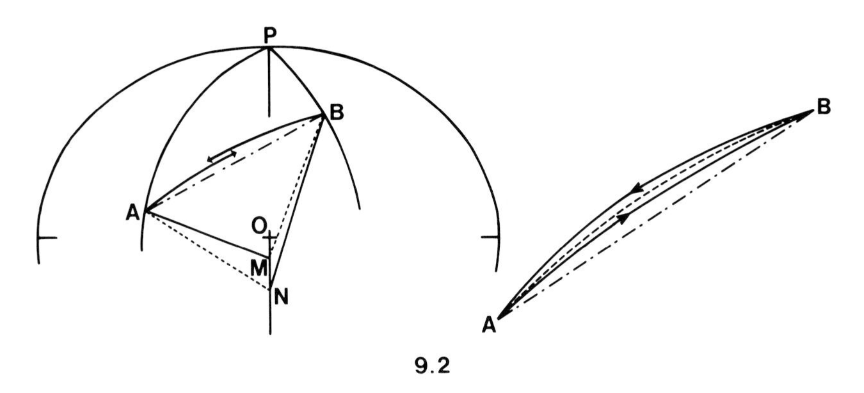

9.2

the curve set out from A. In fact, the two plane sections have the straight chord AB in common, and the relative positions of the two curves will be as shown on fig. 9.2. On the diagrams, the separation of the two plane sections has to be wildly exaggerated. At ordinary survey distances the curves would be hardly distinguishable, at most only a few millimetres apart!

9.3 Curve of alignment

Further consideration of the idea of a surveyor on an ideal spheroidal Earth does, however, lead to the definition of a unique curve on the surface joining A and B. We now suppose that the surveyor goes to places between A and B, and finds points at which the bearings to A and B differ by *exactly* 180°. He would describe this procedure as 'lining himself up' between the two fixed points. He would thus set out the 'curve of alignment' which would obviously start out from A and B in the same directions as the two plane section curves described above. The geometry is indicated on fig. 9.2 with the necessary exaggeration.

The 'alignment' property means that at any point on the curve there is a single normal section plane which contains both points A and B, and therefore contains the straight chord AB. In other words, the curve of alignment is the locus of points at which the spheroid normals actually intersect the chord. Following this definition, the locus may be continued beyond AB as a closed curve round the spheroid; but it is not a plane curve.

9.4 Spheroidal triangle

Three points on a spheroid may be joined by curves of alignment and so we have a practical definition of a spheroidal triangle. The angles at the three points are the angles that would be measured by a surveyor with a theodolite on the spheroid surface.

9.5 Geodesic

Another curve joining two points on a surface is the *geodesic*, the shortest line lying on the surface. A geodesic on a spheroid has the property that the expression $\nu \cos \phi \sin A$ has the same value at all points on it; angle A is the spheroidal bearing. This formula follows from considering the forces on a string held in tension on a perfectly smooth spheroid. The tension is constant and, for equilibrium, its moment about the axis of the spheroid must also be constant.

The geodesic between A and B does not leave these points in the same directions as the curve of alignment, hence the angles between geodesics cannot be directly measured on the surface.

A geodesic may be extended indefinitely in accordance with its formula and it is not a closed curve. For a geodesic starting from a point on the equator with bearing A_0, the constant value of its formula is $a \sin A_0$ and it will go round the spheroid to return to the equator at a point approximately $\pi e^2 \sin A_0$ short of the full 360°. On an Earth-spheroid this is about $(1° \ 12')$ $\sin A_0$.

9.6 Lengths on spheroid

When it comes to considering the lengths of the sides of a spheroidal triangle, it is not necessary in practice to specify a particular curve. At the distances used in practical surveying, the differences between the lengths of the plane sections, geodesic and curve of alignment are quite negligible, far smaller than the probable errors of the most precise survey measurement.

9.7 Plane section ellipse

Because the horizontal angles measured by a surveyor at a point on a spheroid would be angles between plane sections through the normal at the place of observation (as explained in section 9.2), it is of interest to consider in some detail the geometry of vertical plane sections, which in general are ellipses.

Refer again to fig. 9.1 and the plane section at point A where its bearing is A. The section ellipse contains the normal chord AML. If the latitude of A is ϕ the angle AMP is the co-latitude $(90° - \phi)$. It is obvious, geometrically, that somewhere along the section, a point H of highest latitude is reached and there the section will be perpendicular to the meridian plane PH. Furthermore, since a spheroid is symmetrical about any meridian, the plane section is symmetrical about the meridian PH and it follows that the line HMK is the minor axis of the ellipse of section.

Consider the plane section itself on fig. 9.3. Because the line AML is the normal of the spheroid surface, it is normal to any plane vertical section

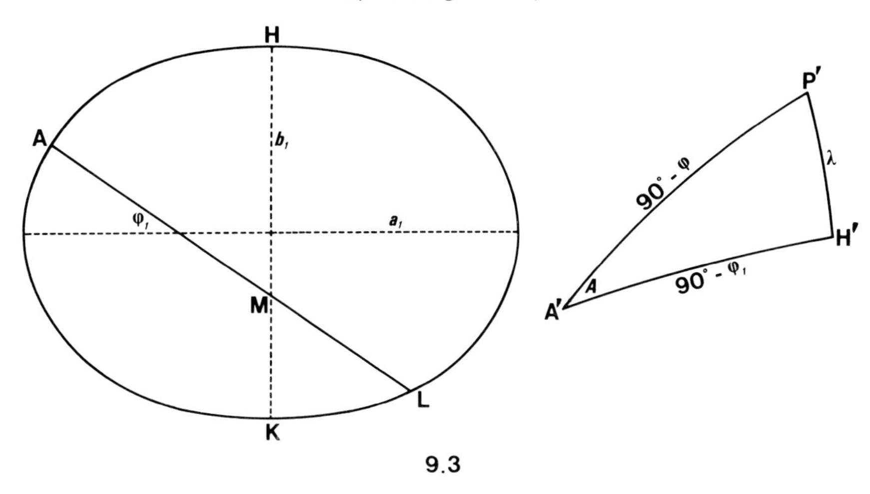

9.3

through A. Thus we see that AM and AL are the normal (ν) and chord of the plane section as well as of the spheroid meridian. The angle marked ϕ_1 is equivalent to latitude on the plane section ellipse, and angle AMH is $(90° - \phi_1)$. Now consider a sphere centred at M on fig. 9.1 and cut by the lines MA, MP and MH. This will generate the spherical triangle A′P′H′ on fig. 9.3 which is right-angled at H′ because the plane section is perpendicular to the plane of meridian PH. (But note that the angle marked λ is not the co-latitude of H because HM is not the normal at H.) For the spherical triangle there are the formulas: $\tan \phi_1 = \tan \phi \sec A$; $\sin \lambda = \cos \phi \sin A$; $\sin \phi = \cos \lambda \sin \phi_1$, etc.

Let the parameters of the plane section ellipse be a_1, b_1, e_1. Then formulas for these can be found by expressing the lengths of AM and AL in terms of both ellipses. Thus:

$$a_1^2(1 - e_1^2 \sin^2 \phi_1)^{-1} = a^2(1 - e^2 \sin^2 \phi)^{-1} \qquad [= \nu^2]$$
$$(1 - e_1^2)(1 - e_1^2 \cos^2 \phi_1)^{-1} = (1 - e^2)(1 - e^2 \cos^2 \phi)^{-1} \qquad [= \text{AL}/2\nu]$$

With some transformations and use of the spherical formulas above we find that:

$$e_1^2 = e^2 \cos^2 \lambda \, (1 - e^2 \sin^2 \lambda)^{-1}$$
$$a_1^2 = a^2(1 - e^2 \sin^2 \lambda - e^2 \sin^2 \phi)(1 - e^2 \sin^2 \lambda)^{-1}(1 - e^2 \sin^2 \phi)^{-1} .$$

As for the radius of curvature ρ_1 along the section, it can be shown from the above formulas, and using formulas proved earlier, that:

$$\frac{1}{\rho_1} = \frac{\cos^2 A}{\rho} + \frac{\sin^2 A}{\nu} .$$

It is proved in the next section that ν is the radius of curvature of the elliptic section when $A = 90°$, so the above relation is just a statement of

Euler's Formula. ρ and ν are the two principal radii of curvature of the spheroid surface.

9.8 Prime vertical section

On putting $A = 90°$ in the formulas of the previous section, the parameters of the prime vertical section can be obtained. In this case, point H is at A so $\lambda = 90° - \phi$, and A is at the pole of the section so $\phi_1 = 90°$. We find that

$$\begin{aligned} e_1^2 &= e^2 \sin^2 \phi(1 - e^2 \cos^2 \phi)^{-1} \\ a_1^2 &= a^2(1 - e^2)(1 - e^2 \cos^2 \phi)^{-1}(1 - e^2 \sin^2 \phi)^{-1} \\ &= \nu^2(1 - e^2)(1 - e^2 \cos^2 \phi)^{-1}. \end{aligned}$$

The chord AL is now the minor axis of the section ellipse so $b_1 = \nu(1 - e^2)(1 - e^2 \cos^2 \phi)^{-1}$ (section 8.9). But the radius of curvature at the pole is a_1^2/b_1 (section 8.5) which is seen to be simply ν. The importance of ν in geodetic computation is due to the fact that it is the radius of curvature of the east–west normal section.

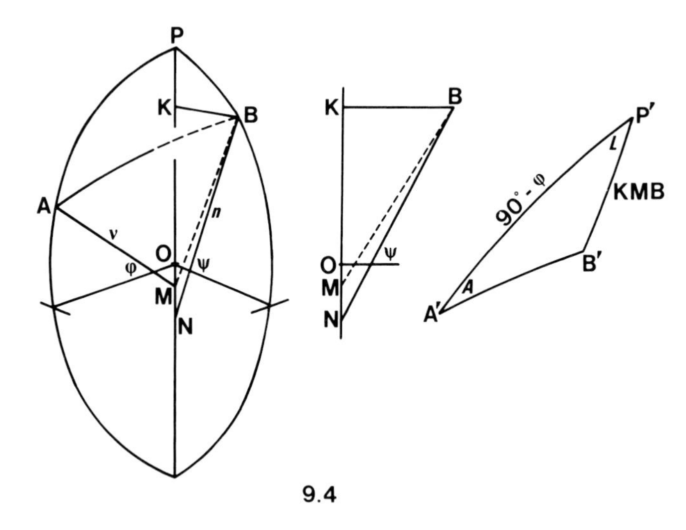

9.4

9.9 Bearings of plane sections

Let the latitudes of two points be ϕ and ψ, with L their difference of longitude. On fig. 9.4, AM and BN are the normals, of length ν and n. By ellipse formulas, OM is $\nu e^2 \sin \phi$ and ON is $ne^2 \sin \psi$, so MN is $e^2(n \sin \phi - \nu \sin \phi)$. BK, the perpendicular from B to the axis of the spheroid, is the radius of the parallel of latitude ψ and has length $n \cos \psi$. KN is $n \sin \psi$. On the diagram that shows the geometry on the meridian of B it is easily seen that:

$$\begin{aligned} \cot(\text{KMB}) &= \text{KM/KB} \\ &= [n \sin \psi - e^2(n \sin \psi - \nu \sin \phi)]/(n \cos \psi). \end{aligned}$$

Now consider a sphere centred at M and cut by the lines MA, MP and MB, generating the spherical triangle shown in fig. 9.4. Angle A, the bearing at A to B, is obtained from the four-part formula: $\sin\phi \cos L = \cos\phi \cot(\mathrm{KMB}) - \sin L \cot A$, which gives:

$$\sin L \cot A = \frac{n \sin\psi \cos\phi - e^2(n \sin\psi - \nu \sin\phi)\cos\phi}{n \cos\psi} - \sin\phi \cos L.$$

To calculate the reverse bearing at B to A, draw the perpendicular from A to the axis and use a spherical triangle centred on N. If B is the reverse bearing measured clockwise from the meridian of B, the angle at B′ in the relevant spherical triangle will be $360° - B$. The formula is found to be:

$$\sin L \cot B = \sin\psi \cos L - \frac{\nu \sin\phi \cos\psi + e^2(n \sin\psi - \nu \sin\phi)\cos\psi}{\nu \cos\phi}.$$

Further reference is made to the above formulas in connection with geodetic computations (chapter 10).

It is interesting to note that these bearings can be calculated by closed formulas without any restrictions as to the flattening of the spheroid or the distance between the points.

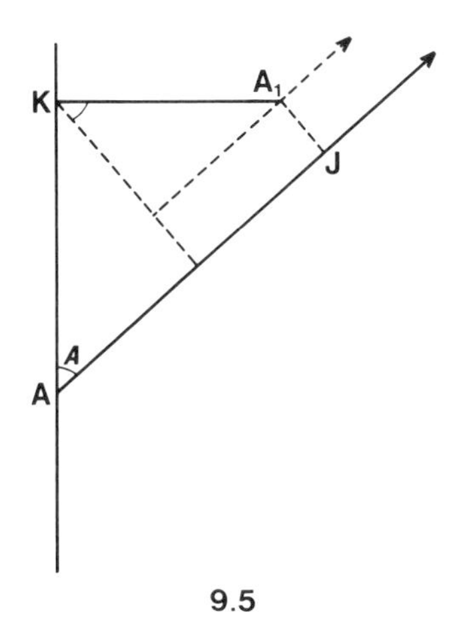

9.5

9.10 Small changes

Point A on the spheroid is at latitude ϕ and the bearing and distance to another point in the survey system are A and s. Suppose that the spheroidal coordinates of A are changed by small amounts $\mathrm{d}\phi$ and $\mathrm{d}L$ to move the point to A_1. Small changes like these may have to be considered in adjustment problems. In linear units the shifts are $AK = \rho\, \mathrm{d}\phi$ and $KA_1 = \nu \cos\phi\, \mathrm{d}L$ ($\mathrm{d}\phi$, $\mathrm{d}L$ in radian). In the geometry of fig. 9.5, it is seen that the effects of both AK and KA_1 are to reduce s, while the effect of AK is to increase the bearing but the effect of KA_1 is to reduce the bearing.

$$AJ = AK \cos A + KA_1 \sin A \text{ and } A_1J = AK \sin A - KA_1 \cos A.$$

So the formulas are strictly:

$$\begin{aligned} ds &= -\rho\, d\phi \cos A - \nu \cos\phi\, dL \sin A \\ dA &= A_1J/s = +(\rho/s)\, d\phi \sin A - (\nu \cos\phi/s)\, dL \cos A. \end{aligned}$$

In using these formulas, the signs of $\cos A$ and $\sin A$ must be correctly maintained.

In practice one would probably want to express $d\phi$ and dL in seconds. The formula for ds is then:

$$-(\rho \cos A/206265)\, d\phi'' - (\nu \cos\phi \sin A/206265)\, dL''$$

and the other formula will give dA in seconds if $d\phi$ and dL are in seconds.

As an example take $\phi = 50°\ 38'\ 51.208''$ and the line AB of bearing 128° 07′ 19.1″ length 36.309 52 km. This is the line used in the computation in section 10.7. Then the changes of length and bearing of AB are given by:

$$\begin{aligned} ds &= +19.075\, d\phi'' - 15.454\, dL'' \text{ metres} \\ dA'' &= +138.10\, d\phi'' + 68.90\, dL''. \end{aligned}$$

CHAPTER 10

Geodetic computation

Some methods are described that have been found suitable for calculating geodetic survey systems on a spheroid surface.

10.1 Geodetic datum

The geometry of a survey system is determined by surveyors by measuring angles and lengths. In most cases some internal adjustments are then applied. Next, the resulting self-consistent figure is to be 'placed' on a suitable spheroid of reference so that the latitudes and longitudes of the survey positions A_0, B_0, etc. of fig. 8.1 can be calculated. In practice the 'placing' is usually done by adopting certain values of spheroidal latitude and longitude at one of the points in the system and the spheroid bearing of one of the lines. This 'datum' might well be the observed astronomical quantities at one of the points.

From then onwards, the calculation of latitudes and longitudes of other points of the survey is a matter of putting the angles and lengths of the geometrical figure into appropriate formulas based on spheroid geometry.

On completion of the calculations, spheroidal latitudes, longitudes and bearings at other points are compared with observed astronomical values. Then it may be considered necessary to revise the datum and the parameters of the spheroid. In this way, spheroids of 'best fit' have been derived for the areas covered by national survey and mapping systems.

10.2 Precision specification

Before considering some methods of computation on a spheroid surface, it will useful to establish specifications of accuracy that will be required in practice. In a conventional triangulation system such as the primary framework for the survey of a country, adjacent points are intervisible and this means distances are mostly in the range 20–60 kilometres. Distances of this order can be measured by E.D.M. with an accuracy of a few parts per million, say 10 centimetres in a line of average length. Angles can be measured to $\frac{1}{2}''$, equivalent to 10 centimetres at a distance of 40 km. In a tightly controlled survey system like triangulation, the adjustment process may be expected to produce a final accuracy somewhat better than the standards quoted above.

We see therefore that practical measuring accuracy is not wasted if distances are expressed to 0.01 metre and angles to 0.1 second. However, to avoid the accumulation of error in extended work, computations may be done to one more decimal place and the final results quoted to the specifications just mentioned.

As regards geodetic coordinates, latitude and longitude, we note that 3 centimetres along a meridian represents 0.001 second of latitude, and this sets a standard conforming with the specifications.

10.3 The classical problem

The calculation of the latitude and longitude of a point from a given bearing and distance has been called the Principal Problem of Geodesy. Several sets of formulas are available for this purpose. Put into symbols, the problem is: At point A latitude ϕ the line AB has bearing A and length s. What is the latitude ψ of B and what is the difference of longitude ω between A and B?

In fig. 10.1, AP is the meridian of A and the broken line through B is part of the parallel of latitude. Whether we regard this diagram as being on a sphere or on a spheroid, it is obvious that at ordinary survey distances the change of position from A to B could be *very nearly* described as a shift northwards of $s \cos A$ followed by a shift eastwards of $s \sin A$. Then the $s \cos A$, on division by a suitable value of meridian radius of curvature, will give *very nearly* the difference of latitude, and the $s \sin A$ will give *very nearly* the difference of longitude on division by the radius of the parallel of latitude of B.

The refinements to produce geodetic accuracy are to be seen in the two sets of formulas to be described below. These, Clarke's formulas and Puissant's formulas, have been widely used in establishing geodetic survey systems.

10.4 Reverse bearings

In addition to the calculation of the latitude and longitude of B, a third formula must be available in practice, because it will usually be necessary to proceed from B to another point C, and the bearing from B to C will then be

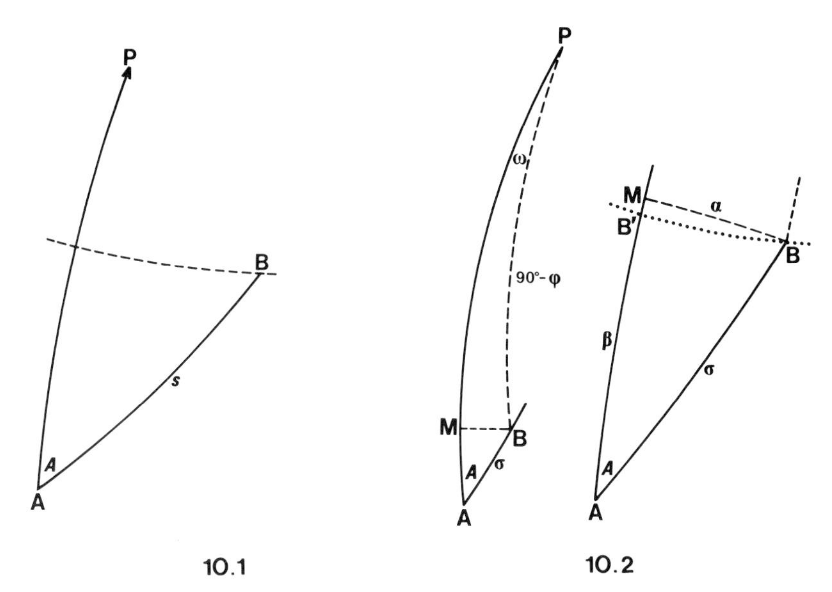

10.1 10.2

required. This will be obtained by calculating the bearing from B to A and adding or subtracting an angle of the survey system.

10.5 Spherical preliminary

In the formulas used by A. R. Clarke for calculating the first principal triangulation of Great Britain, the refinements referred to above are made by two small angles. We first consider the problem on a sphere.

In fig. 10.2, point A has latitude ϕ, the bearing at A to point B is A, and the angle equivalent of the length of AB is σ. We want to calculate ψ, the latitude of B, and ω, the difference of longitude. In practice σ will rarely be as much as $\frac{1}{100}$ radian and ω also will be a small angle of the same order.

For the present we make use of the fact that σ is small. Draw BM perpendicular to meridian PA; denote BM $= \alpha$, AM $= \beta$. By a right-angled triangle formula, $\sin \alpha = \sin A \sin \sigma$, or in series expansions $\alpha - \frac{1}{6}\alpha^3 \ldots = \sin A(\sigma - \frac{1}{6}\sigma^3 \ldots)$. Thus α is approximately $\sigma \sin A$ and this may be substituted into the small term $\frac{1}{6}\alpha^3$ which is then taken across to the right and combined with the $\frac{1}{6}\sigma^3$ term giving, after a little treatment:

$$\alpha = \sigma \sin A - \tfrac{1}{6}\sigma^3 \sin A \cos^2 A \ldots$$

By another formula, $\tan \beta = \cos A \tan \sigma$, and a similar treatment using the series expansion for tangent leads to:

$$\beta = \sigma \cos A + \tfrac{1}{3}\sigma^3 \cos A \sin^2 A \ldots$$

Now if we write $E = \frac{1}{2}\sigma^2 \sin A \cos A$, the cubic terms in the above series can be transformed and the series written as:

$$\alpha = \sigma(\sin A - \tfrac{1}{3}E\cos A \ldots)$$
$$\beta = \sigma(\cos A + \tfrac{2}{3}E\sin A \ldots).$$

But these are also the calculus expansions of:

$$\alpha = \sigma \sin(A - \tfrac{1}{3}E)$$

and $$\beta = \sigma \cos(A - \tfrac{2}{3}E).$$

The latitude of M is $\phi + \beta$, and it is seen that the latitude ψ of B will be slightly less by the small amount MB′. Call this η. BB′ is the parallel through B. Thus the latitude of B is $\phi + \beta - \eta$.

To get η and ω, we go to the right-angled triangle in which the side PM is $90° - (\psi + \eta)$. One of its formulas is $\sin\psi \sec\alpha = \sin(\psi + \eta)$, or in series expansion:

$$\sin\psi\,(1 + \tfrac{1}{2}\alpha^2 \ldots) = \sin\psi + \eta\cos\psi \ldots$$

and cancelling $\sin\psi$ from both sides gives:

$$\eta = \tfrac{1}{2}\alpha^2 \tan\psi.$$

Using the formulas above, η can also be written as $\tfrac{1}{2}\sigma^2\sin^2 A\tan\psi$, and as $E\tan A\tan\psi$. Another formula for η is obtained from triangle PMB and its formula $\cos\omega = \tan\psi\cot(\psi + \eta)$, which by expansion leads to $\eta = \tfrac{1}{2}\omega^2\sin\psi\cos\omega$.

It is obvious that η is a very small angle in our context. The formulas just given must be entered with radian values, and if some realistic numbers are put in it will be seen that η may amount to a few seconds in practice. Hence it will be sufficient to use an approximate value of ψ in calculating η. The latitude of M may be used.

Since the linear lengths of AM and BM are practically $R\sigma\cos A$ and $R\sigma\sin A$, R being the radius of the sphere, the area of the triangle ABM is $\tfrac{1}{2}R^2\sigma^2\sin A\cos A$, and now it will be seen that the formula chosen for E is the spherical excess of the triangle in the situation illustrated in fig. 10.2. But it is to be noted that E must be calculated strictly from the formula and will be negative if A is in the second or fourth quadrant of bearing and the sign must be correctly preserved through subsequent working. (*See* worked example given later.)

Another formula from triangle PMB is $\sin\omega = \sec\psi\sin\alpha$, or, in expansions:

$$\omega - \tfrac{1}{6}\omega^3 \ldots = \alpha\sec\psi - \tfrac{1}{6}\alpha^3\sec\psi \ldots$$

and by a procedure similar to that used before, we get:

$$\omega = \alpha\sec(\psi + \tfrac{1}{3}\eta).$$

So, the required formulas for latitude and longitude are:

$$\psi = \phi + \sigma\cos(A - \tfrac{2}{3}E) - \eta$$
$$\omega = \sigma\sin(A - \tfrac{1}{3}E)\sec(\psi + \tfrac{1}{3}\eta).$$

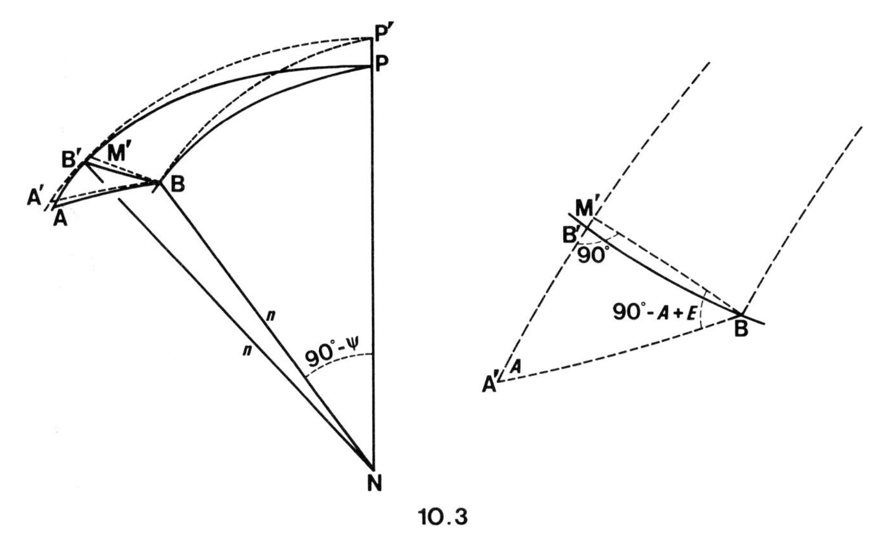

10.3

10.6 Clarke's formulas

In fig. 10.3, point A is on the spheroid at latitude ϕ, the linear distance AB is s and the bearing is A. BB′ is the parallel of latitude ψ and BN, B′N are normals of lenght n. A sphere of radius n centred at N will touch the spheroid along the parallel of B, and it can be shown that the sphere is outside the spheroid elsewhere. All the broken lines on the diagram are on the sphere. A′ is the point on the sphere directly 'above' A. It is now seen that the figure A′M′P′B has the same spherical geometry as fig. 10.2 on the sphere of radius n.

Now it can be shown that, at ordinary geodetic distances, the separation between A and A′ will rarely exceed a metre, so it is evident that the bearing and distance from A′ to B cannot differ significantly from A and s. Thus the 'σ' of fig. 10.2 is now s/n, the 'E' is $\frac{1}{2}(s^2/n^2)\sin A\cos A$, and the *length* of A′M′ in fig. 10.3 is $s\cos(A - \frac{2}{3}E)$. We now require the length of AB′, taken equal to A′B′, to find the difference of spheroid latitude.

Now the angle value of B′M′ is $\frac{1}{2}(s^2/n^2)\sin^2 A\tan\psi$, according to the formula of the previous section, so its linear length is n times this, that is $\frac{1}{2}(s^2/n)\sin^2 A\tan\psi$. By subtraction, therefore, the linear length of AB′ is:

$$s\cos(A - \tfrac{2}{3}E) - \tfrac{1}{2}(s^2/n)\sin^2 A\tan\psi.$$

Now the spheroidal shape is to be taken into account, because to give the difference of spheroidal latitude from A to B′, the above length must be divided by a mean radius of curvature (*see* section 8.11).

Thus the formula for difference of latitude is:

$$\frac{s}{\rho_m}\cos(A - \tfrac{2}{3}E) - \frac{s^2}{2n\rho_m}\sin^2 A\tan\psi.$$

The second term above is the strictly correct expression for the symbol η usually seen in accounts of Clarke's formulas.

Now it is clear that before any values of the small angles E and η can be calculated, it is necessary to find a preliminary, approximately correct, value for ψ, and thence for n. This can be done by calculating $(s \cos A)/\rho$, using any reasonable value for ρ, perhaps its value at ϕ or at some roughly estimated mean latitude of A and B.

As for the longitude difference ω, it is the same on sphere and spheroid, and the formula is:

$$\omega = (s/n) \sin (A - \tfrac{1}{3}E) \sec (\psi + \tfrac{1}{3}\eta).$$

To get the reverse bearing B to A, we look at the angles on the sphere at point B. Since the spherical excess of the triangle A′M′B is E, the angle A′BM′ must be $90° - A + E$. Also we require the angle P′BM′ which we call $90° - \theta$. The side P′B is $90° - \psi$, and a right-angled triangle formula is $\tan \theta = \sin \psi \tan \omega$. By expansion and procedure as used before, this gives:

$$\theta = \omega (\sin \psi + \tfrac{1}{3}\omega^2 \sin \psi \cos^2 \psi \ldots)$$

But it has already been shown that $\eta = \tfrac{1}{2}\omega^2 \sin \psi \cos \psi$, so we can write $\theta = \omega (\sin \psi + \tfrac{2}{3}\eta \cos \psi \ldots)$ which is the expanded form of $\theta = \omega \sin (\psi + \tfrac{2}{3}\eta)$. The reverse bearing B to A is therefore $360° - (90° - A + E) - (90° - \theta)$, that is:

$$B = A + \omega \sin(\psi + \tfrac{2}{3}\eta) - E \pm 180°.$$

Practical working of Clarke's formulas is explained by the numerical example in the next section.

10.7 Clarke's formulas: Example

The formulas as shown above will give angle values in radian. If latitudes, longitudes and bearings are to be expressed in degrees, minutes and seconds, the conversion factor from radian to seconds must be included where appropriate. This number is 206 264.8 . . ., and to save space it will be denoted D. (If results are to be expressed in degrees, the factor is 57.295 777 . . ., or $180/\pi$; and for 'gons' it is 63.661 977 . . . or $200/\pi$.)

On the spheroid defined by a = 6 378.160 km, f = 1/298.25, the point A has latitude ϕ = 50° 38′ 51.208″, the bearing A = 128° 07′ 19.1″ and the distance s is 36.305 92 km. Since A is in the second quadrant, E will be negative and the latitude of B will be less than ϕ. Take $50\tfrac{1}{2}°$ as a rough guess of the mean latitude.

This spheroid has e^2 = 0.006 694 542 . . . We find that ρ at $50\tfrac{1}{2}°$ is 6373.53 km and $D(s/\rho) \cos A$ = − 725″ = − 12′ 05″, so the approximate latitude of B is 50° 26′ 46″, and the mean latitude ϕ_m is about 50° 32′ 48″. At the latitude of B, n is 6390.890 km and at the mean latitude ρ_m is 6373.581 km. Now the small angles can be calculated:

$$E = D(s^2/2n^2)\sin A\cos A = -1.62''$$

and $$\eta = D(s^2/2n\rho_m)\sin^2 A\tan\psi = 2.5008''.$$

On looking at the formulas it is seen that E is connected with angles and bearing, in which an accuracy of 0.1″ is adequate, but η comes into the latitude formula which must be good to 0.001″.

To begin the precise computation, we have:

$$A - \tfrac{1}{3}E = 128°\,07'\,19.6'' \qquad A - \tfrac{2}{3}E = 128°\,07'\,20.2''.$$

Then $$\psi = \phi - 12'\,05.3452'' - 2.5008'' = 50°\,26'\,43.362''$$

Next $$\psi + \tfrac{1}{3}\eta = 50°\,26'\,44.20'' \text{ and } \psi + \tfrac{2}{3}\eta = 50°\,26'\,45.03''$$

Then using the n given above we find that:

$$\omega = 1447.566'' = 24'\,07.566''$$

Finally the reverse bearing:

$$B = 308°\,07'\,19.1'' + 18'\,36.11'' + 1.62'' = 308°\,25'\,56.8''.$$

10.8 Clarke's formulas: Inverse

The so-called 'inverse problem' is the calculation of the precise bearings and distance between two points of given spheroidal latitudes and longitudes. This may have to be done, for instance, in an adjustment of a survey system on the spheroid. One approach to the inverse computation is by way of a rearrangement of Clarke's formulas:

$$s\sin(A - \tfrac{1}{3}E) = \omega n\cos(\psi + \tfrac{1}{3}\eta)$$
$$s\cos(A - \tfrac{2}{3}E) = (\psi - \phi + \eta)\,\rho_m.$$

First ignore the small angles E and η, put radian values of ω and $(\psi - \phi)$ into these formulas, and get approximate values of $s\sin A$ and $s\cos A$. Then:

$$E = \tfrac{1}{2}D\,(s\sin A)(s\cos A)/n^2$$
$$\eta = \tfrac{1}{2}D\,(s\sin A)^2\tan\psi/n\rho_m.$$

Returning to the precise formulas, of which the right-hand sides are now completely known, it is clear that:

$$\frac{\omega n\cos(\psi + \tfrac{1}{3}\eta)}{(\psi - \phi + \eta)\,\rho_m}$$

will be very close to $\tan A$. Calculate this and call it $\tan A'$. It can be proved that a small correction will give:

$$A = A' + \tfrac{1}{3}E(1 + \sin^2 A').$$

The distance s can now be calculated from either of the precise formulas, or both as a check, and the reverse bearing B is obtainable from the third of Clarke's formulas, or by use of the convergence formula explained in section 10.10.

The inverse of the computation in section 10.7 goes as follows:

We have $\omega = 1447.566/D$ radian and $(\psi - \phi) = -727.846/D$ radian. n (at latitude ψ) and ρ_m are calculated. Then, approximately, $s \sin A = 28.5619$ km and $s \cos A = -22.4904$ km, whence $E = -1.62''$ and $\eta = 2.5008''$, as before. Now, going to the complete formulas, we find that $A' = 128° 07' 19.97''$ and the small correction is $-0.87''$, giving $A = 128° 07' 19.1''$. The two calculations of s both give 36.305 92 km.

10.9 Puissant's formulas

This method of dealing with the 'principal problem' has been extensively used in Europe and America.

Refer back to fig. 10.1 and the spherical triangle centred at M. If the side A′B′ (angle AMB) can be calculated, then P′B′ (angle PMB) can be found by a cosine formula.

We are now of course considering situations in which the angle AMB is small – rarely more than half a degree. By finding the parameters a_1, e_1 (section 9.8) of the ellipse of section AB and doing some working with its polar equation (section 8.12), it is shown that the angle AMB is:

$$(s/\nu) + \tfrac{1}{6}e^2(s^3/\nu^3)\cos^2 A \cos^2 \phi \ldots$$

and the second term is quite negligible if s is less than 50 km.

Now the angle PMB is approximately the co-latitude of B and the actual co-latitude, $90° - \psi$, is to be found by subtracting the angle MBN. However, latitudes are required to 0.001″ and with 7-figure logarithms this could not be achieved in the direct computation of angle PMB. Hence the whole process received further mathematical treatment to find formulas for calculating the difference of latitude. Puissant's formula for this is:

$$\psi = \phi + (s/\rho_m)\cos A - \tfrac{1}{2}(s^2/\nu\rho_m)\sin^2 A \tan\phi$$
$$- \tfrac{1}{2}(s^3/\nu^2\rho_m)\sin^2 A \cos A(\tfrac{1}{3} + \tan^2\phi) \ldots$$

The term in s is a first approximation to the difference of latitude and, as with Clarke's formulas, it must be calculated provisionally to get an estimate of the mean latitude ρ_{mi} and then the term in s^2, is a close approximation to η. With the numerical data of section 10.7, the difference of latitude is:

$$-725.3403'' - 2.5188'' + 0.0132'' = -727.846'' \text{ as before.}$$

For the difference of longitude a four-part formula of the spherical triangle is available:

$$\sin A \cot L = \cos\phi \cot(\text{AMB}) - \sin\phi \cos A.$$

Alternatively the formula:

$$\sin L = \sin (s/n) \sin A \sec \psi$$

is sufficiently accurate in most cases. But there may be computing difficulty if the direction of AB happens to be very close to the meridian.

The reverse bearing is obtainable from the convergence formula (*see* section 10.10).

10.10 Dalby's theorem: Convergence

A useful result follows from comparing bearings on sphere and spheroid. A formula in section 9.9 can be written:

$$\sin L \cot A = \cos \phi \tan \psi - \sin \phi \cos L - e^2 (n \sin \psi - \nu \sin \phi) \cos \phi / n \cos \psi.$$

On putting $e^2 = 0$, we get the formula for calculating the bearing A_0 when the two points are on a sphere and have the same latitudes ϕ, ψ, and difference of longitude L:

$$\sin L \cot A_0 = \cos \phi \tan \psi - \sin \phi \cos L$$

which is simply one of the four-part formulas of the spherical triangle.

Hence, $\sin L (\cot A_0 - \cot A) = e^2 (n \sin \psi - \nu \sin \phi) \cos \phi / n \cos \psi.$

The difference between A and A_0 is small and if we put $A = A_0 + \theta$, then $\cot A = \cot A_0 - \theta \operatorname{cosec}^2 A_0 \ldots$, hence:

$$\theta \sin L = e^2 (n \sin \psi - \nu \sin \phi) \cos \phi \sin^2 A_0 / n \cos \psi.$$

But by sine formula, $\cos \phi \sin A_0 = \cos \psi \sin B_0$, and so:

$$\theta \sin L = e^2 (n \sin \psi - \nu \sin \phi) \sin A_0 \sin B_0 / n.$$

Evidently if the same process is carried out for the difference $B - B_0$, the formula will be the same except for having ν in the denominator instead of n. Therefore the differences between spheroid and spherical bearings at A and at B will be very nearly equal.

Consider the line that has been computed in previous sections. By appropriate spherical formulas it is found that the bearings, if the points were on a sphere, would be $A_0 = 128° \, 11' \, 51.9''$ and $B_0 = 308° \, 30' \, 29.6''$ differing from the spheroid bearings by $4' \, 32.8''$ in each case.

Another way to express this result is to say that the change of bearing from A to B, or *convergence* as it has been called, is practically the same, whether the points are on sphere or on spheroid. This is Dalby's theorem. It follows that the convergence can be calculated by the (exact) spherical formula:

$$\tan \tfrac{1}{2}\Delta A = \tan \tfrac{1}{2}L \sin \tfrac{1}{2}(\psi + \phi) \sec \tfrac{1}{2}(\psi - \phi).$$

Then $$B = A + \Delta A \pm 180°.$$

Indeed, the simple formula $\Delta A = L \sin \phi_m$ is sufficiently accurate in most cases. It gives the same value 18′ 37.7″ in the worked example.

10.11 Some small corrections

As indicated in section 8.1, the purpose of a geodetic survey operation is to determine the geometry of the figure formed by the points A_0, B_0, etc. on a surface of reference. Since observations cannot be made at these points, but must be made at the accessible points A, B, there arises the question of what 'corrections' ought to be applied to the observable quantities in order to obtain measures of the corresponding quantities on the reference surface.

Measured lengths are customarily 'reduced' to horizontal and further 'reduced' to sea-level by one part per million for each 6.4 metres (or 21 feet) of height. Ideally the 'height' should be the distance above the spheroid. If the spheroid first adopted is shown to be a poor fit to sea level, some reassessment, height corrections and readjustment of the system may be necessary.

The theodolite measures angles around the vertical, that is the direction of gravity at each point. The angles of the spheroid figure are those that would be measured round the normals AA_0, BB_0, etc. Divergences between vertical and normal are 'deviations of the vertical' which can be revealed by making astronomical observations. When the deviations are known, corrections to horizontal angles can be calculated. The treatment is the same as for the effects of mislevelment of the theodolite as described in section 7.11.

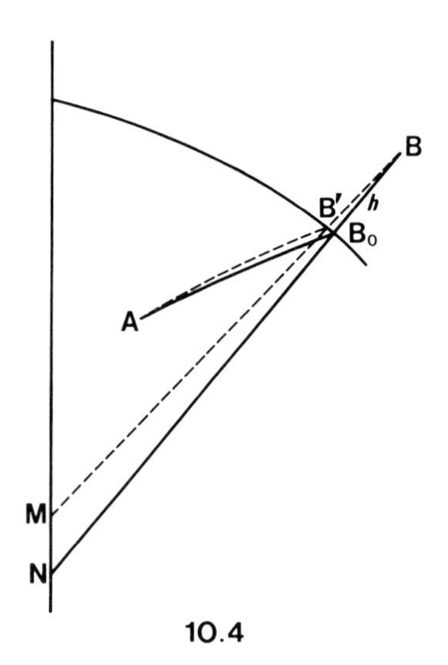

10.4

Another effect on horizontal angles is directly due to the spheroidal shape. Referring to fig. 10.4, suppose that the observer at point A sets his telescope on point B which is at height h above the spheroid. If the telescope is lowered, the line of sight will move down the line BM, point M being on the normal at A. Thus the horizontal reading of the instrument will refer to point B′, not to B_0 which is the spheroid point representing the position of B. The horizontal reading should be corrected by the angle subtended by $B'B_0$ at point A. The correction is very small, indeed usually negligible in practice, and the simple formula $\frac{1}{2}(h/a) \sin 2A \cos^2 \phi$ (radian) is sufficient. ϕ and A can be approximate mean values for the line. For $h = 1000$ m and the line at bearing 45° in middle latitude, the correction is less than 0.1″.

CHAPTER 11

Global geodesy

By use of artificial satellites, geodetic measurement has now become truly global. Points separated by hundreds or thousands of kilometres may be connected in spatial geometrical figures to an accuracy comparable with that of ground measurement over the same distances. It has become possible to determine the shape of the geoid over the oceans.

11.1 3-D systems

In a conventional geodetic survey done by triangulation, trilateration or traversing, adjacent points have to be interversible, hence the lines sighted or measured will have lengths mostly less than 60 km and rarely as much as 100 km. The primary purpose of such surveys, for mapping or engineering use, is the determination of the relative positions of points on the Earth's surface.

It has been found convenient to use a geometrical reference surface (spheroid) in the treatment of such surveys. Relative horizontal positions are expressed as spheroidal latitudes and longitudes, and the reference surface can be represented on a plane by means of a map projection. The 'third dimension', height, is in practice referred not to the spheroid but to a 'natural' datum, usually mean sea level.

Global geodetic measurement has led to the use of three-dimensional coordinate systems. There are two possibilities: one is to adopt a spheroid as before and take the third coordinate to be the distance above the spheroid measured along the extended normal; the other is to go to a system of three rectangular coordinates suitably defined as to origin and axes, thus dispensing with any form of reference surface.

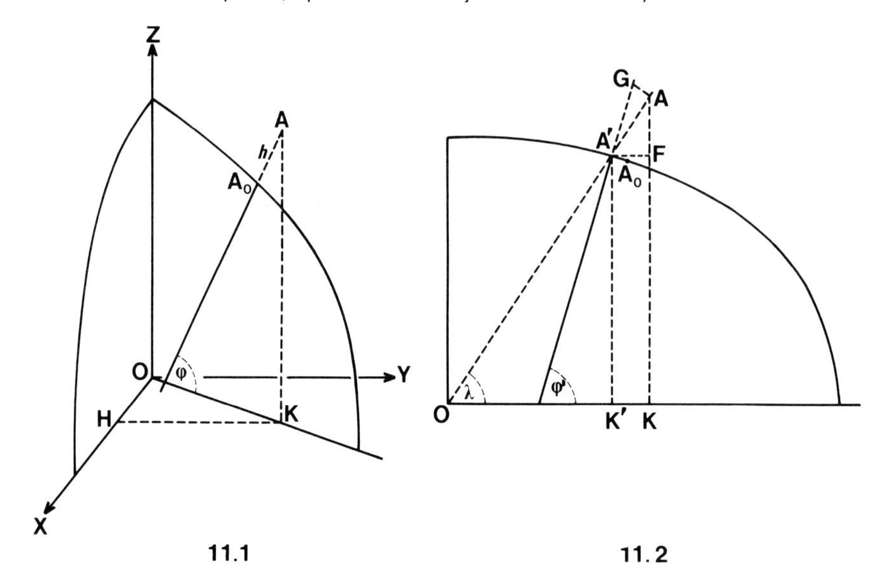

11.1

11.2

11.2 Spheroidal 3-D

In fig. 11.1 the point A_0 is at latitude ϕ on the spheroid and point A is at a distance h along the normal at A_0. With reference to a specified zero meridian ZOX, the longitude of A_0 is L, angle KOH, and so the three coordinates giving the position of A are L, ϕ, h.

These coordinates can be converted to a purely rectangular system with origin at the centre of the spheroid. Having regard to the formulas given in section 8.6, it is clear that:

$$\mathrm{OK} = (\nu + h)\cos\phi \text{ and } \mathrm{KA} = (\nu(1 - e^2) + h)\sin\phi.$$

It is usual practice to take the axis of the spheroid as OZ with the OX axis in an adopted zero longitude meridian and the OY axis making a right-handed set as indicated on fig. 11.1. In this system therefore the three rectangular coordinates referred to origin O would be:

$$\begin{aligned} X &= \mathrm{OH} = (\nu + h)\cos\phi\cos L \\ Y &= \mathrm{HK} = (\nu + h)\cos\phi\sin L \\ Z &= \mathrm{KA} = (\nu(1 - e^2) + h)\sin\phi. \end{aligned}$$

As an example let A_0 be at latitude 50° on the spheroid defined by $a =$ 6 378 160 m, $f = 1/298.25$. Then it is found that $\nu = 6\,390\,725.396$ m, and on adding a height $h = 1000$ m the position of A in its own meridian is given by OK = 4 108 521.890 m and KA = 4 863 572.059 m. The X and Y coordinates are (OK cos L) and (OK sin L) respectively.

Conversion the other way – given X, Y and Z, to find h and the latitude and longitude of A_0 – is not quite straightforward; but advantage is taken of the fact that h will in practice be very small in comparison with the dimensions of the spheroid. Finding the longitude is of course a simple matter: $\tan L = Y/X$; and the radial length OK is $(X^2 + Y^2)^{\frac{1}{2}}$. Fig. 11.2 shows the meridian of A. The join AO meets the spheroid at A′ and clearly $\tan \lambda = Z/\text{OK}$. But λ is the geocentric latitude of A′ and by ellipse formulas the latitude of A′ is given by:

$$\tan \phi' = \tan \lambda/(1 - e^2) = Z/(1 - e^2)\text{OK}.$$

Hence the ellipse coordinates of A′, OK′ and K′A′, can be calculated. Obviously ϕ' will be very close to ϕ in practice. The normal at A′ is produced and AG is perpendicular to it. A′F and FA are the differences of coordinates between A′ and A.

In the numerical example OK will be as given above, and ϕ' is found to be 50.000 029 74° = 50° 00′ 00.1071″. The coordinates of A′ are OK′ = 4 107 876.568 m and K′A′ = 4 862 808.136 m, so the differences are A′F = 645.322 m and FA = 763.921 m.

It is easily seen that A′G = (A′F $\cos \phi'$ + FA $\sin \phi'$) and GA = (A′F $\sin \phi'$ − FA $\cos \phi'$), respectively 1000.003 m and 3.307 m. For all practical purposes A′G can be taken as the height h, while the 3.307 divided by the appropriate radius of curvature, strictly $(\rho + \text{h})$, gives the difference of latitude from A′ to A_0; it is 0.1071″ as expected.

11.3 Rectangular 3-D

Orbits of artificial satellites are ‘tracked’ from ground stations and, in turn, the positions of other ground stations, or ships at sea, can be found by observations to the satellites. This process determines the relative positions of points in ‘Earth–space’, and the use of a simple three-dimensional co-ordinate frame of reference is a direct way for describing the positions. No geometry of sphere or spheroid is involved.

A 3-D coordinate system could have its origin at any specified point and its axes in any specified, mutually perpendicular, directions. However, it is usual to adopt axes that are ‘Earth centred’ in the same sort of way as indicated in the previous section. In fact, the positions of axes and origin are located by the giving of rectangular coordinates to points on the Earth’s surface.

By connecting geodetic survey systems to available satellite observation stations, the spheroids used for mapping may be joined in a common frame of position, and geodesists may study the geometry of the Earth–as–a–whole.

Part III Map projection

CHAPTER 12

Curved surface to plane map

We discuss here some general considerations on the relationship between a spherical or spheroidal surface and its representation on to a plane.

12.1 Map projection

The construction of a map requires the transformation of points on a spherical or spheroidal surface to points on the plane surface of the map. Because in practice the features indicated on the map are expected to bear some resemblance, as regards shape and relative positions, to the corresponding features on the Earth, a basically geometrical approach is made to designing a suitable transformation system or *map projection*. Although a projection must be expressible as a pair of formulas for converting latitudes and longitudes into a plane coordinate system, it is usual to describe a projection by explaining the geometrical structure of the network or *graticule* of lines on the map that will represent the network of meridians and parallels on the sphere or spheroid.

Very few projections are projections in the geometrical sense, and some of them cannot be drawn by ruler–and–compasses methods. Drawing must usually be done by plotting calculated coordinates of a series of points and joining by smooth curves. However, the computer-controlled plotting table is now available, not only for ruler–and–compasses jobs, but for plotting a complete map from stored positional information and programmed projection formulas.

12.2 Projection properties

Following from its particular method of construction, each projection will have certain properties of interest to the cartographer and surveyor. Angles, distances, areas, etc. on the sphere or spheroid will have their counterparts on the projection. Some will be represented exactly and others with discrepancies (often described in the literature as projection *errors*). In studying a map projection one must be careful to compare projection quantities with the quantities they actually represent on the curved surface. It is impossible for any plane projection to show distances correctly at the same scale throughout, and this leads to the question as to what is the precise meaning of the *scale* that is ascribed to a map.

12.3 Scale of a projection

For the construction of a series of topographical or cadastral maps, the surface to be 'projected' is a portion of a spheroid of specified dimensions (a and f given); for an atlas map, when there is no need to bother about the ellipticity of the Earth, the projected surface is a notional spherical 'Earth' having some reasonable radius such as 6370 km. In either case there is no difficulty in imagining a *model* of the surface reduced to any required scale, e.g. 1/20000000 for an atlas map or 1/50000 for a topographical series. The surface of the model may then be represented on a plane by a projection designed to preserve closely the dimensions of the model, and the scale of the model may be quoted as the (nominal) scale of the map.

12.4 Scale at a point

Some projections are deliberately constructed so that certain lines, for example all the meridians, are represented exactly at the stated scale of the map. The description *equidistant* may be applied to such constructions.

Again, there may be lines along which the scale, while not being true, is the same along the whole line, and then the actual scale may be calculated by comparing the projection length of any portion of the line with its true length on the scaled model.

Otherwise, the scale along most lines in most projections is not uniform, and it is not very helpful to compare a finite portion of such a line with its true model length. However, one may compare 'infinitesimally short' lengths on projection and model, and thus obtain the *scale at a point* along a line. The ratio, or *scale factor*, which may be expressed symbolically as $(\mathrm{d}s_1/\mathrm{d}s)$, is a fundamental parameter in projection systems, and obviously it may be studied by applying calculus methods to the projection formulas.

12.5 Orthomorphic, authalic, or otherwise

Although the linear scale cannot be constant over a projection, it is possible to preserve angles or areas exactly, but not both simultaneously. Angles are correctly represented on *orthomorphic* (also called *conformal*) projections. That is, if two lines on the sphere or spheroid cut at a certain angle, the lines that represent them on the projection will cut at exactly the same angle. Well-known examples are the stereographic, mercator and transverse mercator projections, the last-named being very widely used for topographical map series. *Authalic* (or *equivalent* or *equal–area*) projections preserve area. That is, equal areas on the map represent equal areas on the reference surface. This means however that shapes may be considerably distorted, because if there is an exaggeration of linear scale in one direction, there must be a compensating reduction in the cross direction and some consequent distortion of angles. It is advisable to use equal–area projections for atlas maps giving certain types of information where relative areas are significant.

It is perhaps worth mentioning that if the scale of an authalic map is expressed as a ratio of areas, e.g. 1 square kilometre to 25 square millimetres, such a ratio is exact over the whole extent of the projection and not nominal as a linear scale ratio must be – 1/200 000 in this case.

There are also many projections having neither of the properties mentioned above, but based on simple geometrical constructions. These can be seen in atlases, and a few including the Cassini and polyconic projections, have been used for topographical map series.

CHAPTER 13

Some simple projections of a sphere

Methods for studying the properties of a projection can be easily understood by applying them to plane representations of a spherical surface. The surface to be represented is a spherical 'model' of the Earth, as mentioned in section 12.3: the essential properties of a projection are expressed in its relationship to this model. In this chapter a few simple and well-known projections of a sphere are described.

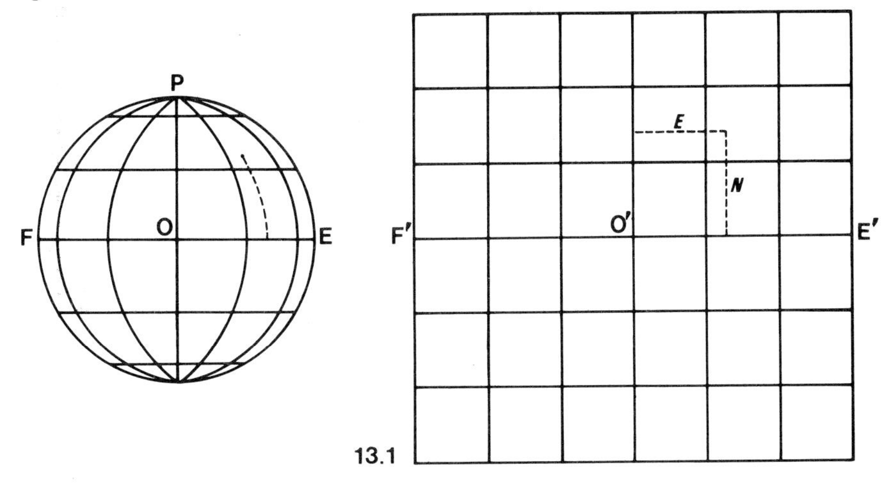

13.1

13.1 Cylindrical equidistant

Perhaps the simplest conceivable geometry for representing the meridians and parallels is a 'squared paper' structure as indicated in fig. 13.1. The circular diagram (which is itself a projection of a hemisphere) shows meridians and parallels at 30° intervals. The equator FOE is represented by

F′O′E′ 'true to scale', that is of length πR, where R is the radius of the sphere model. All the meridians are also of this same length and true to scale.

Obviously, scale along the parallels is increasingly exaggerated with increasing latitude. On the hemisphere the parallel of latitude ϕ has length $\pi R \cos \phi$ and the projection length, regardless of latitude, is πR, so the exaggeration is by factor sec ϕ which is, for instance, 2 at latitude 60°. In high latitudes distortion of shape will be severe; this projection should not be extended to the Poles except for solely diagrammatic representations of information.

Taking point O′ as an origin of rectangular coordinates, shown as E and N, easting and northing, the geometry can be expressed by the formulas $E = RL$ and $N = R\phi$ (L and R in radian).

This projection has also the names 'simple cylindrical' and 'plate Carrée'.

The true to scale equator F′O′E′ is the basis for several other projections having properties depending on other ways of spacing out the parallels.

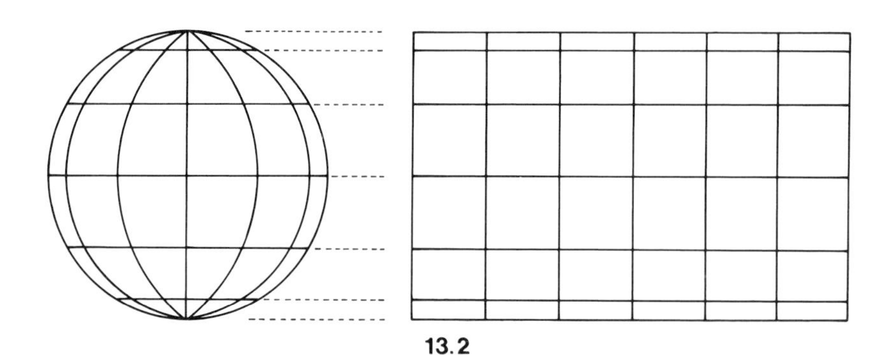

13.2

13.2 Cylindrical equal–area

For instance, if the spacing is as indicated on fig. 13.2, the result is an equal–area (authalic) representation; this follows from the equality of areas of the sphere and the enveloping cylinder. Consider a narrow zone between latitudes ϕ and ϕ + dϕ. It has width R dϕ and length all round the sphere $2\pi R \cos \phi$, hence area $2\pi R^2 \cos \phi \, d\phi$. The corresponding strip round the cylinder has length $2\pi R$ but its width is $R \, d\phi \cos \phi$ because the element of meridian, R dϕ, is inclined at angle ϕ to the cylinder surface, so again the area of the strip is $2\pi R^2 \cos \phi \, d\phi$.

The coordinate formulas are obviously $E = RL$ and $N = R \sin \phi$.

Clearly the distortion of shape on this projection in high latitudes will be even worse that in the equidistant projection. The east–west exaggeration of scale by factor sec ϕ must here be compensated by north–south reduction factor, cos ϕ. A small area at latitude 60° will be not only stretched to double its 'width' but also compressed to half its 'depth'.

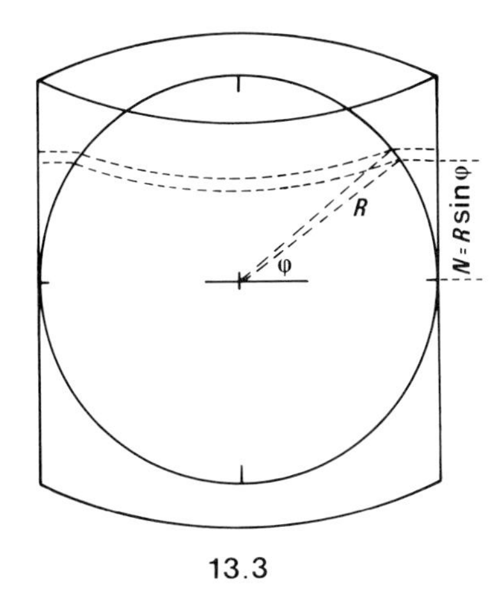

13.3

13.3 Scale at a point

In the two projections described above, the scale along each parallel is *uniform,* so the scale factor is obtainable by comparing any portion of the parallel with the corresponding portion on the sphere. In the equal–area projection, the scale along a meridian is not uniform. So we have to consider the idea of 'scale at a point'. Obviously this is to be found by comparing an 'infinitesimally short' line on the sphere model with its representation on the projection. The element $R\,d\phi$ on the model meridian is to be compared with dN on the projection. In the equal–area projection, dN is $R \cos\phi\, d\phi$, so the meridian scale factor is $\cos\phi$, the inverse of $\sec\phi$ as it should be.

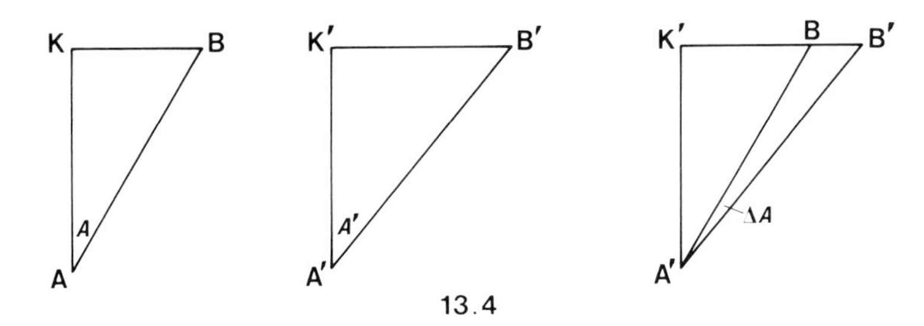

13.4

13.4 Infinitesimal triangles

The triangle AKB in fig. 13.4 is composed of infinitesimally short lengths on the sphere, AK on a meridian and KB on the parallel. In the cylindrical equidistant projection, AK is represented by A′K′ of the same length but KB is represented by K′B′ of length KB $\sec\phi$, as shown in the second diagram. The two triangles are combined in the third diagram.

These infinitesimal triangles can be used to give complete information about the properties of the projection. Consider a numerical example. Let AK be 1 infinitesimal unit and let angle KAB be 30°, the bearing of the line AB. Then KB is tan 30° that is 0.57735 and AB is sec 30°, that is 1.15470. Suppose the latitude is 45°. Then K′B′ is KB sec 45°, that is 0.81650. A′K′ = AK = 1, so tan A' is 0.81650 and A' is 39° 14′. This is the projection bearing that represents a true bearing 30° at latitude 45°. The bearing distortion 9° 14′ shows as angle BKB′.

If similar calculations are done with bearing KAC = 60°, it is found that the projection bearing is 67° 47.5′. Thus the true angle BAC = 30° on the sphere shows as 28° 33.5′ on the projection.

To complete the information obtainable from fig. 13.4 we note that the element of length AB is represented by A′B′, so the scale factor in this direction is the ratio:

$$A'B'/AB = \sec A'/\sec A = \cos A/\cos A'$$

as can be seen by applying a sine formula to the triangle BA′B′. At latitude 45° bearing 30°, the factor is 1.11803 and at bearing 60° it is 1.32288.

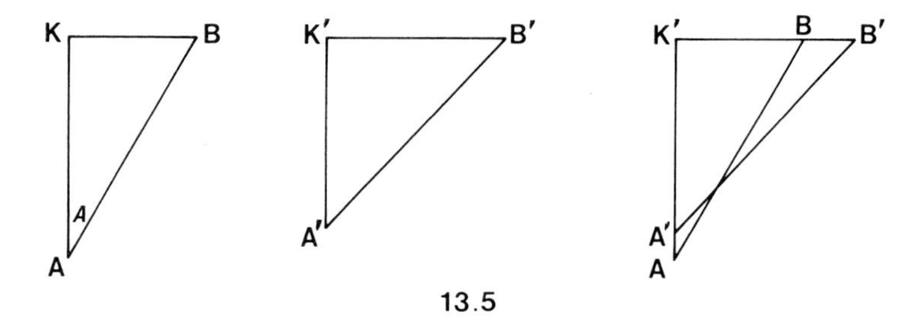

13.5

Now apply the foregoing treatment to the equal–area projection, with the infinitesimal triangles in fig. 13.5. In this case the meridian element AK is to be reduced by the multiplier cos ϕ, so A′K′ is 0.70711. K′B′ is as before and the projection bearing is found to be 49° 06.5′. The scale factor for direction AB is now $\cos \phi \sec A'/\sec A$ which amounts to 0.93541 for bearing 30°.

Incidentally, it is obvious that for this projection there will be a direction through A for which the scale is 'true' – scale factor = 1. Some simple trigonometry shows that this is so when $\tan A = \cos \phi$; at latitude 45° bearing 35° 16′, for example.

13.5 Cylindrical orthomorphic: Mercator

From the 2:1 stretch of shape at latitude 60° in the equidistant projection, it was necessary to go to 4:1 to produce equal–area. What about going the other way and having 2:1 exaggeration north–south as well as east–west? The infinitesimal triangles on the sphere and projection will have the same angles; bearings, and therefore angles, on the projection will be 'true'. This

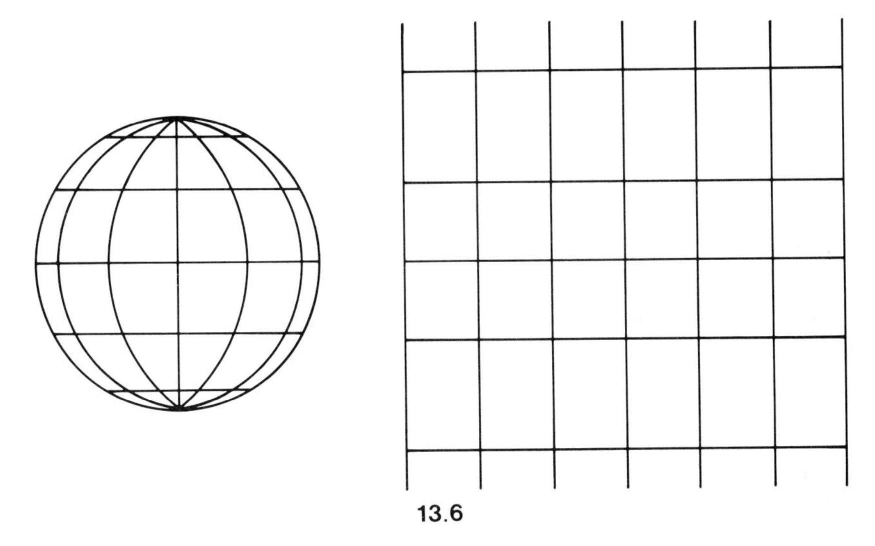

13.6

is orthomorphism. So, we want the scale factor on the meridian to be sec ϕ as along the parallel. In calculus terms this is $dN = R\, d\phi \sec \phi$. The solution of this equation is:

$$N = R \ln \tan (45^\circ + \tfrac{1}{2}\phi) = R \ln (\sec \phi + \tan \phi).$$

(N.B. 'ln' is the 'natural' logarithm to base 2.71828... and is equal to the ordinary (base 10) logarithm multiplied by 2.302585...) This projection is partially shown in fig. 13.6: it 'goes to infinity' towards the poles. It is the famous mercator projection much used as an aid to navigation, and now in its transverse form much used for mapping (*see* section 15.5).

If a straight line is drawn on a map that has the meridians represented by parallel straight lines it necessarily cuts all the meridians at the same angle. But the mercator projection is orthomorphic, so in this case the straight line represents a line on the sphere cutting all the meridians at this same angle, i.e. bearing. Such a line on the sphere is called a *loxodrome* or *rhumb-line*. So the navigator measures the bearing on his mercator chart and sails so as to keep always heading on that bearing. This will not be the shortest route but it will get him to where he wants to go.

Three cylindrical projections have been described. Other systems for spacing out the parallels may be, and have been, invented *ad lib,* and their properties of scale and angle representation worked out. No further examples are considered here.

13.6 Zenithal projections

A different kind of structure for projections is suggested by the meridians radiating from a pole. The simplest forms of zenithal projection have the parallels of latitude represented by concentric circles and the meridians as straight lines radiating from the common centre at true longitudinal angles.

As with the cylindrical projections, the system of spacing out the parallels will determine the geometrical properties of the projection. In any case it is clear that the scale along each parallel will be *uniform*.

In studying the properties of these projections it is mathematically convenient to use the co-latitude $\lambda = (90° - \phi)$ for designating the parallels. If r is the projection radius of parallel λ, each projection has its particular formula for r in terms of λ.

Parallel λ on the sphere has linear radius $R \sin \lambda$, so the scale factor along the parallel is $r/(R \sin \lambda)$. If $d\lambda$ and dL are infinitesimal changes of co-latitude and longitude, the infinitesimal triangle on the sphere has sides $R\, d\lambda$ and $R \sin \lambda\, dL$. These are represented on the projection as dr and $r\, dL$, as in fig. 13.7.

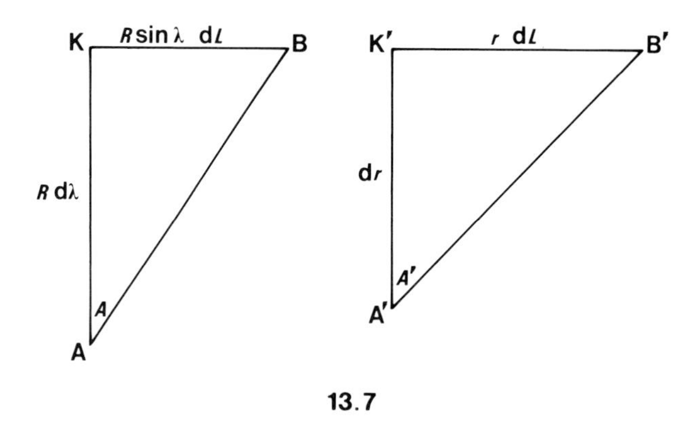

13.7

13.7 Zenithal equidistant

Reckoning distance from the pole, r is $R\lambda$ and the scale factor along the parallels is therefore $\lambda/\sin \lambda$. This is always greater than 1. Along the circle representing the equator, for instance, the factor is $\frac{1}{2}\pi = 1.5708$. Obviously the factor will increase rapidly beyond $\lambda = 90°$. This projection for a hemisphere is shown on fig. 13.8.

The infinitesimal triangle used before – at latitude 45° with AK = 1 and AB on bearing 30° – transforms to the projection as follows: The scale along the parallel is $(\pi/4)/(1/\sqrt{2}) = 1.11072$, so K′B′ is $(1.11072)(0.57735) = 0.64127$, and the projection bearing A' is 32° 40.25′. The scale factor along A′B′ is 1.0288.

13.8 Zenithal equal-area

To produce the zenithal authalic projection we refer back to fig. 13.3 and note that the spherical 'cap' from the pole to co-latitude λ corresponds with a

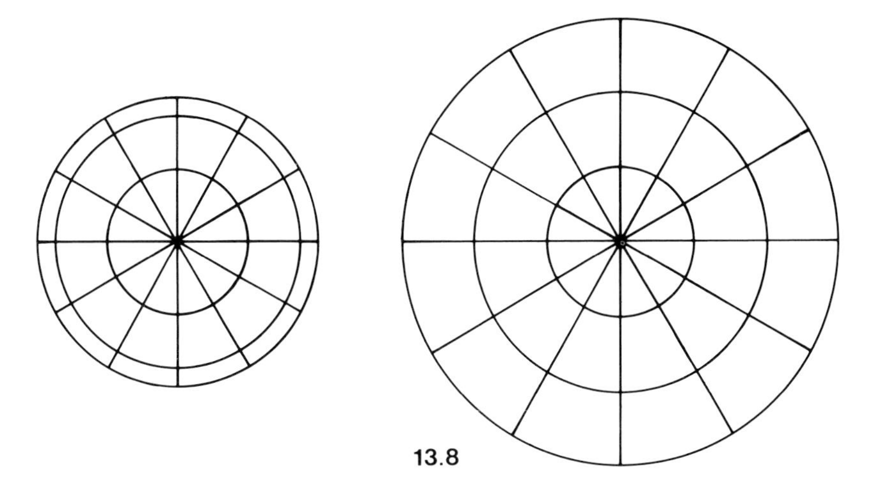

13.8

strip of length $(R - R\cos\lambda)$ on the cylinder, so its area is $2\pi R^2(1 - \cos\lambda)$ which is $4\pi R^2 \sin^2 \frac{1}{2}\lambda$. This is the area of a circle of radius $r = 2R \sin \frac{1}{2}\lambda$. It is seen that this is the length of the straight chord from the pole to the parallel λ, and thus the projection can be drawn by the simple construction indicated in fig. 13.9.

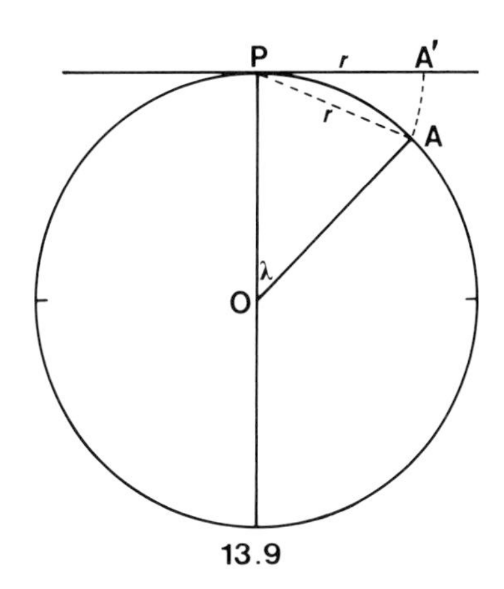

13.9

The scale factor along parallel λ is $2\sin\frac{1}{2}\lambda/\sin\lambda$, that is $\sec\frac{1}{2}\lambda$, so the factor along the meridian is $\cos\frac{1}{2}\lambda$, as is of course seen by differentiating the formula for r.

Taking the infinitesimal triangle AKB as before, we find that in this projection $A'K' = \cos 22.5° = 0.92388$, and $K'B' = \tan 30° \sec 22.5° = 0.62492$. Whence $\tan A'$ is 0.67641 and $A' = 34° \; 04.5'$. The scale factor along A′B′ is 0.96595. In general, the scale factor is exactly 1 when $\tan A = \cos\frac{1}{2}\lambda$, in this example on bearing 42° 44′.

It is of some interest to derive the formulas for this projection without reference to the area of the spherical 'cap', by writing the requirement that the areas of the triangles in fig. 13.7 shall be equal, that is:

$$\tfrac{1}{2}r\,\mathrm{d}r\,\mathrm{d}L = \tfrac{1}{2}R^2 \sin\lambda\,\mathrm{d}\lambda\,\mathrm{d}L.$$

On multiplying both sides by 4 and omitting the dL, this differential equation integrates to the complete solution $r^2 = -2R^2 \cos\lambda + C$. The 'constant of integration' is really necessary here; without it r is imaginary! If C is $2R^2$, we get the formula given above. If C is greater than $2R^2$, then r has a positive value for $\lambda = 0$, the pole is represented by a circle, and there is a 'hole' in the middle of the projection! Such a construction is of course strictly equal–area, but hardly a valuable contribution to cartography.

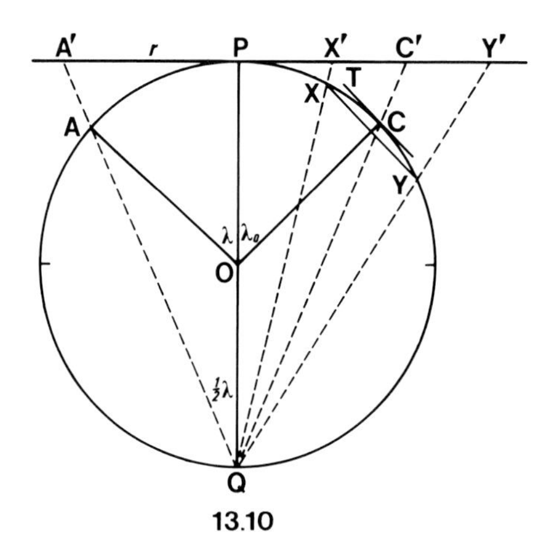

13.10

13.9 Zenithal orthomorphic: Stereographic

We now come to the well-known stereographic projection which is the orthomorphic member of the zenithal class.Referring again to fig. 13.7, the requirement that the elementary triangles shall be the same shape ($A' = A$) is $\mathrm{d}r/(r\,\mathrm{d}L) = R\,\mathrm{d}\lambda/(R \sin\lambda\,\mathrm{d}L)$, for which the solution is $\ln r = \ln \tan \tfrac{1}{2}\lambda +$ constant. That is $r = C \tan \tfrac{1}{2}\lambda$. The factor C is really arbitrary because orthomorphism is an angle relationship which does not determine a scale. This projection is a truly geometrical projection and is usually illustrated by the construction shown in fig. 13.10, giving the formula $r = 2R \tan \tfrac{1}{2}\lambda$, in which case the scale factor is $\sec^2 \tfrac{1}{2}\lambda$ by differentiation.

An interesting and useful property of the stereographic projection is that any circle, great or small, on the surface of the sphere is represented by a circle on the plane. In fig. 13.10, XY represents the plane of a small circle that has its centre at C. All the projecting lines from Q to points on the circle form an elliptic cone which cuts the projection plane at points X′Y′, etc., and the axis of the cone is QCC′ because XC = CY, and therefore the angles XQC and YQC are equal. Now if the co-latitude of C is λ_0 it is seen that QC′

meets the projection plane at angle $90° - \frac{1}{2}\lambda_0$. Moreover the radius OC is perpendicular to the tangent at C and angle OCQ is also $\frac{1}{2}\lambda_0$. So angle OCT is also $90° - \frac{1}{2}\lambda_0$. The chord XY is parallel to the tangent CT. Thus the small circle and its projection are sections of the cone by planes inclined to the cone axis at the same angle. So the two sections are the same shape and, since XY is circular, its projection is also circular. Note however that C′ is not the centre of the projection circle.

13.10 Gnomonic projection

Another zenithal projection with a property that can be useful is the gnomonic, constructed by geometrical projection from the centre of the sphere. The formula is obviously $r = R \tan \lambda$. Any great circle on the sphere projects as a straight line, namely the intersection of the plane of the great circle with the plane of the projection. The scale factor is $\sec \lambda$ along parallels and $\sec^2 \lambda$ along meridians. This projection 'goes to infinity' as λ approaches 90°.

13.11 Orthographic projection

This is simply the zenithal construction centred on the pole with all the parallels 'true' size, hence true to scale. Our view of a distant spherical object such as a planet is practically an example of this geometry. The smaller diagram in fig. 13.8 is an orthographic projection of a polar hemisphere.

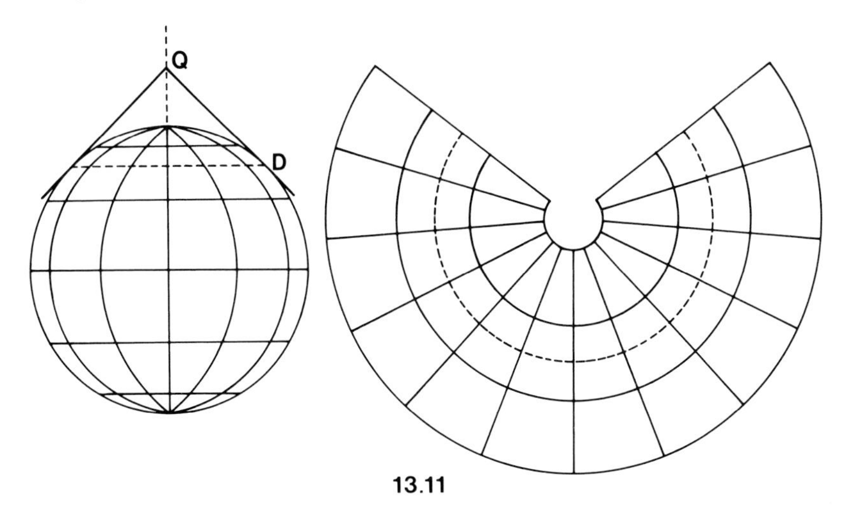

13.11

13.12 Conical projections

Conical projections have straight meridians radiating from a point, but the angles between them are proportionally less than actual differences of longitude. This proportion is a fraction k called the *constant of the cone* and is usually in the range of $\frac{1}{2}$ to $\frac{3}{4}$.

Conical projections are often introduced by considering a cone touching the sphere as shown in fig. 13.11. The cone is then to be slit along one meridian line and opened out flat. If it touches the sphere along parallel of latitude ϕ_0, this parallel will be the same length, $2\pi R \cos \phi_0$, on the cone but its radius will be QD which is $R \cot \phi_0$, hence the projection will cover an angle of only $2\pi \sin \phi_0$ round Q. So in this case the constant of the cone is $\sin \phi_0$; or $\cos \lambda_0$ if co-latitude is used.

As with the cylindrical and zenithal projections, the system of spacing out the parallels is at choice, and the factor k provides another 'degree of freedom' for the cartographer. This makes possible the mapping of a considerable range of latitude without excessive distortion.

It is not necessary to start with the 'cone fitting' geometry and any construction based on the definition first given above, is conical. Some writers include even more constructions in the conical category.

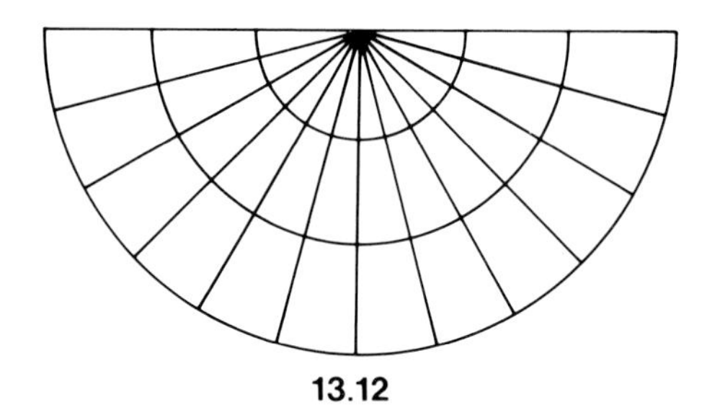

13.12

13.13 Conical equidistant projections

To construct the equidistant projection on the cone of fig. 13.11 after it has been opened out, concentric circular arcs are drawn at 'true' distances apart, and it will be seen that the pole is represented by an arc of radius $(R \tan \lambda_0 - R\lambda_0)$. The resulting projection of the hemisphere, on the cone touching the sphere at latitude 45°, is shown on the figure. The parallel 45° true to scale is called a *standard parallel*, and it is obvious that the scale factor along all other parallels is greater than unity. This projection is usually given the name 'simple conical'.

The equidistance property may start at the pole and the k is at choice. A simple construction for a hemisphere is shown in fig. 13.12; k is $\frac{1}{2}$ and $r = R\lambda$. The scale factor along parallel λ is $\pi R\lambda/(2\pi R \sin \lambda)$, that is $\frac{1}{2}\lambda \operatorname{cosec} \lambda$. Close to the pole this is 0.5 and it increases to $\pi/4 = 0.7854$ at the equator.

This construction might well be extended to 30° beyond the equator ($\lambda = \frac{2}{3}\pi$) where the scale factor is about 1.2. Some patient trial and error work will show that the scale is true at about $\lambda = 108° 36'$.

In any conical projection as defined, the scale along parallel λ is true if $kr = R \sin \lambda$. Equidistant projections are included in the general formula $r = r_0 + R\lambda$, and r_0 is the radius of the arc representing the pole. The parameters r_0 and k are at choice, so it is possible to have true scale along two specified parallels. The conditions are:

$$k(r_0 + R\lambda_1) = R \sin \lambda_1 \text{ and } k(r_0 + R\lambda_2) = R \sin \lambda_2.$$

By subtraction we find that $k = (\sin \lambda_2 - \sin \lambda_1)/(\lambda_2 - \lambda_1)$ and some further work gives:

$$r_0 = R(\lambda_2 \sin \lambda_1 - \lambda_1 \sin \lambda_2)/(\sin \lambda_2 - \sin \lambda_1).$$

For example, if the standard parallels are to be 30° and 60°, it is found that $k = 0.69906$ and $r_0 = 0.19165\, R$.

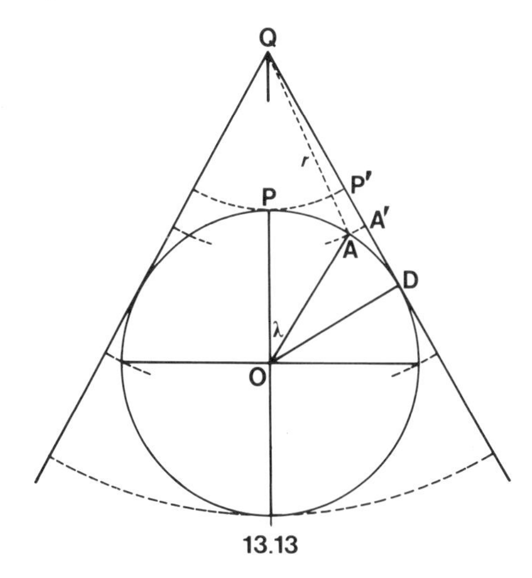

13.13

13.14 Conical authalic projections

A conical equal–area projection on a tangent cone may be produced by the construction indicated in fig. 13.13. The projection radius of each parallel is equal to its distance from the apex of the cone. If λ_0 (angle POD) is the co-latitude of the touching parallel then OQ is $R \sec \lambda_0$. By a formula for the plane triangle QOA we have:

$$(QA)^2 = r^2 = R^2 + R^2 \sec^2 \lambda_0 - 2R^2 \sec \lambda_0 \cos \lambda.$$

The pole is represented by a circular arc of radius QP $= (R \sec \lambda_0 - R)$. For the touching cone, $k = \cos \lambda_0$ (section 13.12). Hence the area of the projection between the pole and parallel λ is $\cos \lambda_0$ times the difference of the areas of circles of radii QP and QA, that is:

$$\cos \lambda_0[\pi(R^2 + R^2 \sec^2 \lambda_2 - 2R^2 \sec \lambda_0 \cos \lambda) - \pi(R^2 + R^2 \sec^2 \lambda_0 - 2R^2 \sec \lambda_0)]$$

which simplifies down to $2\pi R^2(1 - \cos \lambda)$. But this is the area of the spherical cap from the pole to co-latitude λ (section 13.3).

The above geometry is interesting in that the constructions for the zenithal and cylindrical equal–area projections are simply its special cases, when Q is at P and at infinity respectively.

For the derivation of general formulas of conical projection the elementary infinitesimal triangles are similar to those for zenithal projection, the only difference being that the side K′B′ (fig. 13.7) is $kr\,\mathrm{d}L$. The equal–area property requires $kr\,\mathrm{d}r\,\mathrm{d}L = R^2 \sin \lambda\,\mathrm{d}\lambda\,\mathrm{d}L$, which has the general solution $kr^2 = C - 2R^2 \cos \lambda$ with 'disposable' parameters k and C.

The parameters can be found for the conical authalic projection with the two standard parallels. Scale is true on parallel λ if $kr = R \sin \lambda$. Rewrite the general formula as $kC = 2k R^2 \cos \lambda + k^2r^2$, substitute $R \sin \lambda$ for kr, and then the two conditions for true scale are:

$$kC - 2kR^2 \cos \lambda_1 = R^2 \sin^2 \lambda_1$$

and

$$kC - 2kR^2 \cos \lambda_2 = R^2 \sin^2 \lambda_2.$$

These give the explicit expressions for k and C:

$$k = \frac{\sin^2 \lambda_2 - \sin^2 \lambda_1}{2(\cos \lambda_1 - \cos \lambda_2)} = \tfrac{1}{2}(\cos \lambda_1 + \cos \lambda_2)$$

$$C = R^2(1 + \cos \lambda_1 \cos \lambda_2)/k.$$

Projections worked out in this kind of way would normally be used for mapping a limited range of latitude, and then the standard parallels can be chosen so that the scale factors differ very little from unity anywhere. As an example take $\lambda_1 = 30°$ and $\lambda_2 = 40°$; then it is found that $k = 0.8160$ and $C = 2.0384\,R^2$. The projection radii are given by $r^2 = 2.4509\,R^2(1.0192 - \cos \lambda)$. The scale factors on parallels of co-latitudes 25°, 35°, 45°, are 1.0157, 0.9962, 1.0093, respectively, and of course the reciprocals of these numbers along the meridians.

By some rather more sophisticated mathematical treatment, given the limited range of latitude to be covered, it is possible to locate the standard parallels so that the greatest scale 'error' is at a minimum. This sort of refinement is more of a mathematical amusement than a serious problem in cartography.

13.15 Conical orthomorphic

A general formula for conical orthomorphic projections is obtained by requiring the infinitesimal triangles to be similar, that is:

$$\mathrm{d}r/(kr\,\mathrm{d}L) = R\,\mathrm{d}\lambda/(R \sin \lambda\,\mathrm{d}L) \text{ or } \mathrm{d}r/r = k \operatorname{cosec} \lambda\,\mathrm{d}\lambda.$$

This has the general solution:

$$\ln r = \text{constant} + k \ln \tan \tfrac{1}{2}\lambda, \text{ or } r = C(\tan \tfrac{1}{2}\lambda)^k.$$

To find the parameters for this projection with two specified standard parallels, the true scale condition $kr = R \sin \lambda$ is expressed logarithmically as:

$$\ln r = \ln R + \ln \sin \lambda - \ln k$$

and this is combined with the logarithmic general formula $\ln r = \ln C + k \ln \tan \tfrac{1}{2}\lambda$, written in terms of the two specified latitudes. Treatment similar to that done for the equal–area projection leads to:

$$k = \frac{\ln \sin \lambda_2 - \ln \sin \lambda_1}{\ln \tan \tfrac{1}{2}\lambda_2 - \ln \tan \tfrac{1}{2}\lambda_1}$$

then follows:

$$\begin{aligned} \ln k + \ln C - \ln R &= \ln(kC/R) \\ &= \frac{\ln \sin \lambda_1 \ln \tan \tfrac{1}{2}\lambda_2 - \ln \sin \lambda_2 \ln \tan \tfrac{1}{2}\lambda_1}{\ln \tan \tfrac{1}{2}\lambda_1 - \ln \tan \tfrac{1}{2}\lambda_1}. \end{aligned}$$

Taking λ_1 and λ_2 as in the previous section we find that $k = 0.8209$, $kC/R = 1.4726$, $C = 1.7954\,R$. The scale factor is:

$$\begin{aligned} kr/(R \sin \lambda) &= kC(\tan \tfrac{1}{2}\lambda)^k/(R \sin \lambda) \\ &= 1.4726\,(\tan \tfrac{1}{2}\lambda)^k \operatorname{cosec} \lambda. \end{aligned}$$

At co-latitudes 25°, 35°, 45°, the factors are 1.0128, 0.9962, 1.0107, respectively.

13.16 Secant conical projections

An obvious elementary geometry to produce a conical projection with two standard parallels is a cone cutting the sphere along the two parallels. When this is opened out flat these parallels are true to scale but they will be less than the true distance apart. Some scheme for spacing out the other parallels must be found; there are several possibilities.

13.17 The conical class

The cylindrical, zenithal and conical systems of projection described above may be regarded as a single class, in which the first two types are the limit cases of the conical system when the semi-angle of the cone is 0° and 90°. The meridians are straight lines that are concurrent (at 'infinity' in the cylindrical systems) and the parallels of latitude are concentric circles, or parallel straight lines, that cut the meridians at right angles.

A few projections not coming under the above definitions will complete this chapter.

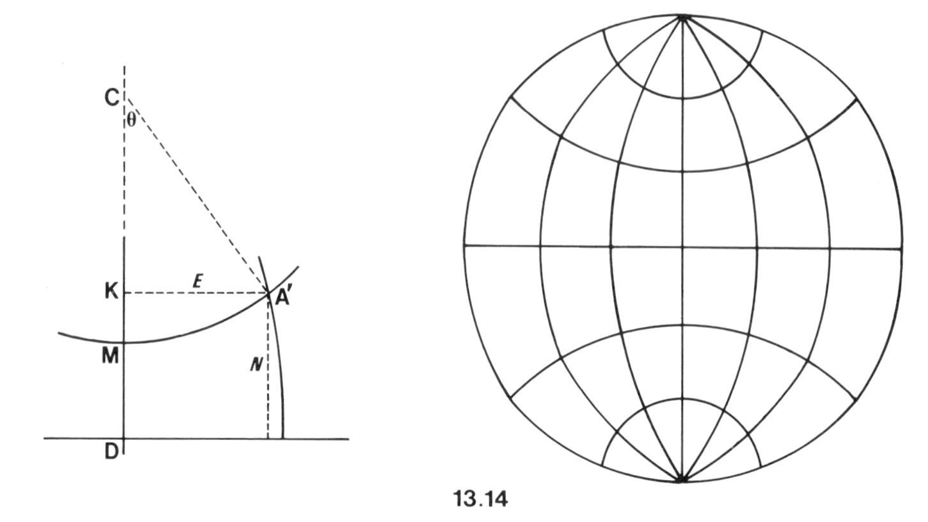

13.14

13.18 Polyconic projection

This method of representation has a very simple geometry but no other notable virtue. A selected meridian through the area to be mapped is represented by a straight line divided true to scale. Then each parallel is represented by a circular arc centred on this meridian and passing through the appropriate point on it. The radius of the circle is $R \cot \phi$ (or $R \tan \lambda$, *see* fig. 13.11). It will be found that the parallels diverge as shown on fig. 13.14. All the parallels are divided true to scale and the points on each meridian are joined by smooth curves; there is no simple mathematical formula for the meridian curves. The geometry for graphical construction is shown on the figure. CD is the straight central meridian, point D being on the equator. Then DM is $R\phi$ and MC is $R \cot \phi$. C is a different point for each parallel. At longitude L on the sphere, the length along the parallel ϕ is $R \cos \phi L$ from the central meridian, so this is also the curve length MA′ on the projection. Hence the angle θ is $L \sin \phi$. Then MK is $R \cot \phi(1 - \cos \theta)$ and KA′, the easting coordinate of A′, is $R \cot \phi \sin \theta$. By joining the straight chord MA′ it is seen that the angle MA′K is $\frac{1}{2}\theta$, so MK can be expressed also as $E \tan \frac{1}{2}\theta$.

Thus the simple formulas for the projection coordinates are $E = R \cot \phi \sin \theta$ and $N = R\phi + E \tan \frac{1}{2}\theta$, where $\theta = L \sin \phi$. For marking off the point A′ graphically, one may use the chord length MA′ which is $2R \cot \phi \sin \frac{1}{2}\theta$ (*see* section 15.1).

13.19 Bonne projections

A Bonne projection starts similarly with a central straight meridian true to scale. A selected parallel ϕ_0 is represented by a circular arc of radius $R \cot \phi_0$

centred at the appropriate point C. And now the rest of the parallels are *concentric* circles at the true distances apart. These are then divided true to scale and the meridians joined by curves. Hence the radius for parallel ϕ is evidently:

$$r = R \cot \phi_0 - R(\phi - \phi_0).$$

The construction is very similar to that for the polyconic. Angle θ is $(LR \cos \phi)/r$ and then $E = r \sin \theta$ and $N = R\phi + E \tan \frac{1}{2}\theta$.

The form of a Bonne projection depends on the initially chosen parallel, but all Bonne projections are authalic. Consider an 'infinitesimally small rectangle' bounded by two adjacent parallels and two adjacent meridians on the sphere. In a Bonne construction the parallels are true distance apart and the lengths along the parallels are also true. The rectangle on the sphere will generally become a parallelogram but the areas will be equal.

If the initial parallel is the equator all parallels are represented by straight lines and the meridians are cosine curves. This is the Sanson–Flamsteed projection. If the initial parallel is the pole then $r = R\lambda$. This is the Werner projection.

These two projections are illustrated on fig. 13.15 by drawings for a hemisphere.

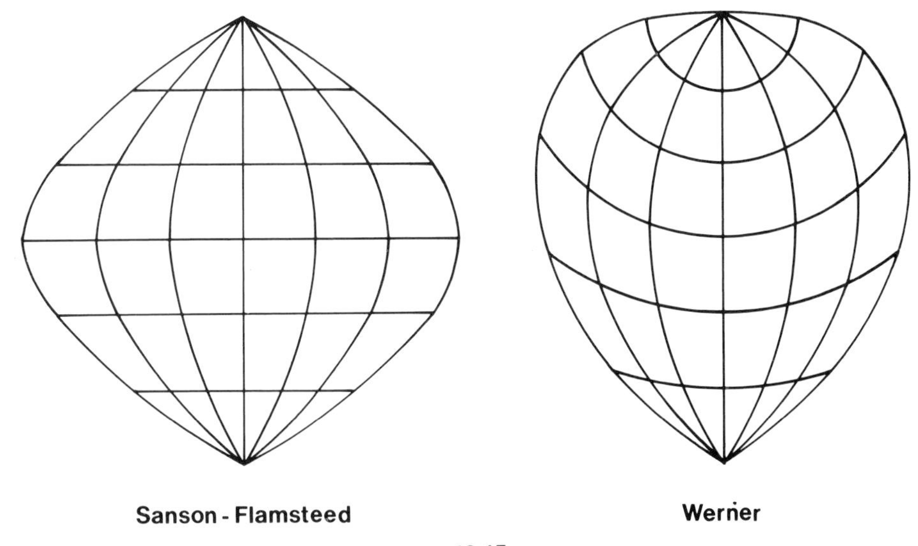

13.15

CHAPTER 14

Transverse and oblique positions

It is to be noted that all the projections described so far are related directly to the meridians and parallels on the sphere; they are centred round the polar axis. This relationship is sometimes called *normal aspect.* Now the surface of a sphere is the same shape all over. Therefore any construction for the projection of a sphere can be similarly related to any diameter of the sphere and will have the same intrinsic geometrical properties in relation to this diameter as the 'normal aspect' projection has in relation to the polar axis. A few examples will best illustrate the methods to be applied.

The descriptions of these projections must take into account the requirement that whatever the position of the projection geometry, the cartographer's aim will be to construct the network, or graticule, representing the ordinary meridians and parallels of the Earth model.

14.1 Transverse cylindrical projections

For instance, the structure of any cylindrical projection may be related to a diameter of the equator, as on fig. 14.1, and then the meridian QGP will be geometrically its 'equator' and we must imagine a 'transverse' pattern of meridians and parallels shown as broken lines on the figure.

Let A be a point of latitude ϕ, longitude L reckoned from meridian PGQ. EAD is the transverse 'meridian' through A, giving rise to the right-angled spherical triangle PAD with elements as shown on the figure. The side p and the angle ϕ'(GD), can be calculated by standard formulas. ϕ' is the angle GED. We can now think of p as the 'latitude' in the transverse system with PGQ as 'equator' and ϕ' as 'longitude' referred to the transverse 'meridian' FGE.

If we now plot the linear lengths of GD and DA as rectangular coordinates

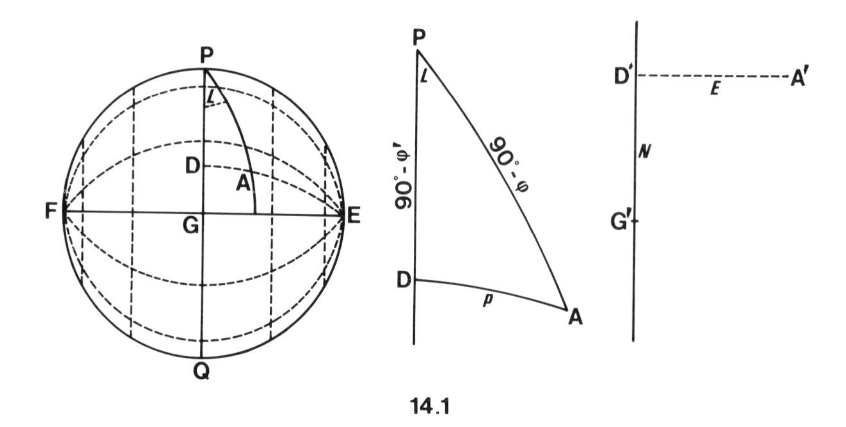

14.1

$N = R\phi'$, $E = Rp$, we get the transverse version of the structure of the cylindrical equidistant projection. This is basically the geometry of the Cassini projection which has been much used in the past for topographical mapping.

Similarly, plotting E as $R \sin p$, will construct the transverse cylindrical equal–area projection, and plotting E as $R \ln (\sec p + \tan p)$ will give the transverse mercator projection.

In all cases the graticule of the real meridians and parallels must be constructed by calculating rectangular coordinates of series of points and joining up the curves. Scale and angle properties will be related to p and ϕ' in the same way as they are related to ϕ and L in the normal aspect. For instance, in the Cassini projection the scale is true along the transverse meridians such as D'A' and the scale factor is $\sec p$ along the transverse parallels that are small circles centred on point E.

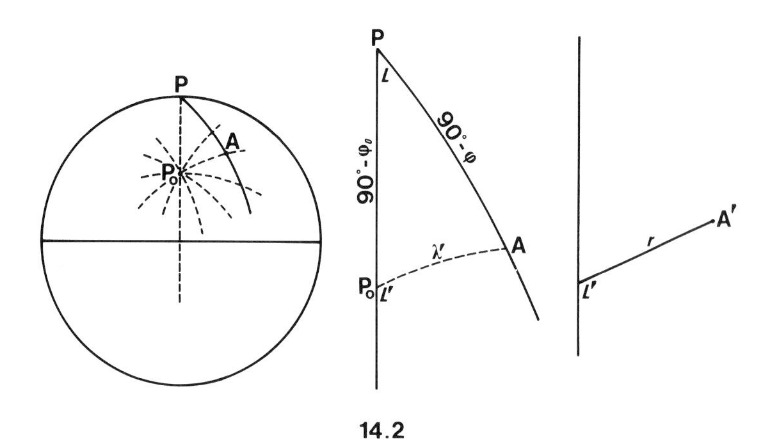

14.2

14.2 Oblique stereographic projection

Now consider the construction of a stereographic projection centred round a point of latitude ϕ_0. We now have to imagine a pattern of 'oblique' meridians radiating from this point P_0, and parallel small circles centred round P_0, as in fig. 14.2. The spherical triangle PAP_0 has the elements shown, and standard formulas can be used to calculate L' and λ' in terms of ϕ and L. Taking $r = 2R \tan \frac{1}{2}\lambda'$ and plotting A′ by the polar system (rL') as indicated, will produce the desired projection. The scale factor at any point is $\sec^2 \frac{1}{2}\lambda'$.

Of course this projection retains the important property of representing circles on the sphere by circles on the plane.

14.3 Equatorial zenithal authalic projection

Point E_0 is on the equator. Point A has latitude ϕ and its longitude is L reckoned from the meridian of E_0. The spherical triangle E_0PA has side $E_0P = 90°$. Side E_0A is σ and angle PE_0A is L'. By a cosine formula $\cos \sigma = \cos \phi \cos L$ and by a four-part formula $\tan L' = \cot \phi \sin L$.

If points along meridians and parallels are plotted in the polar system $(2R \sin \frac{1}{2}\sigma, L')$ the projection mentioned above will be constructed.

CHAPTER 15

Spheroid projection

There is no point in taking account of the ellipticity of the Earth when constructing an atlas map. Maps on such a small scale are graphic presentations of information about the Earth's surface and Man's activities on it, and about location, and transport. The cartographer's problem is to choose projection systems that suit the areas to be covered and the type of information to be presented. In particular, an equal–area projection should be used if comparisons between areas of countries or regions are significant.

On the other hand, in mapping a small portion of the Earth on a 'topographic' scale such as 1/50000, the distortions in any reasonable projection will be graphically insignificant. The surveyor tends to think of such a projection primarily as a plane coordinate system into which he may have to fit his survey measurements. Projection distortions may be invisible on a map but they exist in the mathematical formulas. Survey work connected to points in a plane coordinate system must be treated in such a way as to take account of the projection structure and produce correct coordinates of any new points surveyed.

There is also no point in systematically describing classes of projections of a spheroid. For one reason, the number of projections in actual use for large-scale map series is very small. For another reason, a spheroid is *not* the same shape all over, so a projection structure in the 'normal aspect' cannot simply be moved to another position and retain exactly the same geometry; basic ideas can be 'moved over' but precise formulas are peculiar to each individual projection.

15.1 Polyconic

We start with the polyconic projection, for which the construction, in normal aspect to the spheroid, is the same as for the sphere. Distances along the central meridian are given by the arc length formula in section 8.10, and each parallel is drawn with the radius $\nu \cot \phi$ (*see* fig. 15.1). The parallels are true to scale: at longitude L the length of the parallel from the central meridian is $\nu \cos \phi L$, so the angle θ (fig. 13.14) is $L \sin \phi$. It is easily seen that the plane coordinate formulas are:

$$E = \nu \cot \phi \sin \theta \text{ and } N = m + E \tan \tfrac{1}{2}\theta.$$

This projection is suitable for an area of limited extent in longitude. It was adopted in 1909 as the basis of the World Map Series at scale 1/1 million, for which there is a separate projection for each 6° of longitude. Some modifications have been made to the strict polyconic structure to minimise scale 'errors' and simplify the geometry. For instance, meridians are drawn as straight lines, so that each sheet fits the adjoining sheets on all four sides.

The polyconic is not authalic or orthomorphic, and the modern tendency is towards the use of orthomorphic coordinate systems.

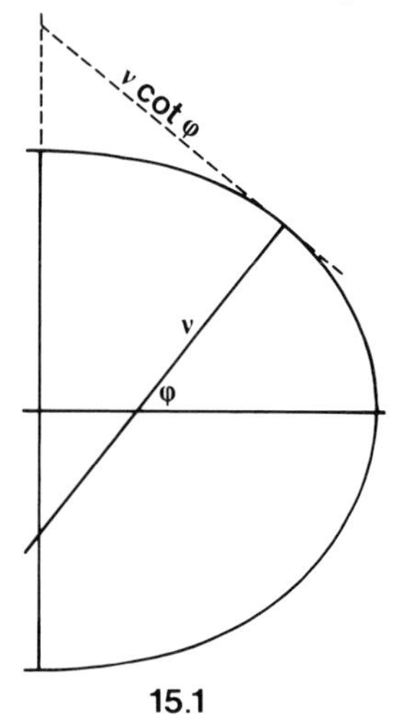

15.1

15.2 Bonne

The geometry is the same as for a sphere, starting with an initial parallel of radius $\nu_0 \cot \phi_0$, and the other parallels concentric at the meridional distances apart. By construction the equal–area property is maintained.

15.3 Mercator

This projection could be used for a large-scale map system in an area extending along the equator.

In a cylindrical projection of a spheroid with the equator at true scale, the parallel at latitude ϕ of length $2\pi \nu \cos \phi$, is represented by a length $2\pi a$, so the scale factor is:

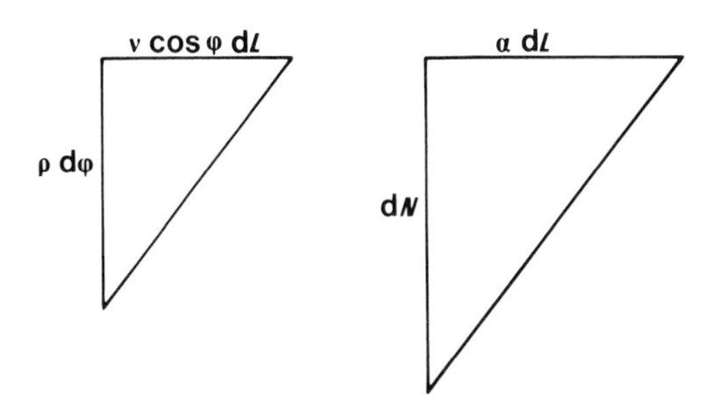

15.2

$$a/(\nu \cos \phi) = (1 - e^2 \sin^2 \phi)^{\frac{1}{2}} \sec \phi.$$

The infinitesimal triangles for small changes $d\phi$ and dL, are as in fig. 15.2. For orthomorphism:

$$dN/a = (\rho/\nu) \sec \phi \, d\phi = \sec \phi \, d\phi \, (1 - e^2)(1 - e^2 \sin^2 \phi)^{-1}.$$

By writing $(1 - e^2)$ as $(1 - e^2 \sin^2 \phi - e^2 \cos^2 \phi)$, the equation becomes:

$$dN = a \sec \phi \, d\phi - ae^2 \cos \phi \, d\phi/(1 - e^2 \sin^2 \phi).$$

The substitution, $e \sin \phi = \sin v$, then gives:

$$dN = a \sec \phi \, d\phi - ae \sec v \, dv$$

whence

$$N = a \ln (\sec \phi + \tan \phi) - ae \ln (\sec v + \tan v).$$

A constant of integration is not necessary because the above formula gives $N = 0$ at $\phi = 0°$.

For example, at $\phi = 3°$ we find that $v = 0.245\,349\,15°$ and:

$$N/a = 0.052\,383\,82 - 0.000\,350\,37 = 0.052\,033\,45$$

using the spheroid dimensions as in previous examples. An alternative writing of the formula is:

$$N = a \ln \tan (45° + \tfrac{1}{2}\phi) - ae \ln \tan (45° + \tfrac{1}{2}v).$$

15.4 Polar stereographic

For a zenithal projection in normal aspect to the spheroid, the infinitesimal triangles are as in fig. 15.3; r is the projection radius for the parallel of co-latitude λ and ρ and ν must be expressed in terms of λ. For orthomorphism:

$$dr/r = (\rho/\nu) \operatorname{cosec} \lambda \, d\lambda$$
$$= \operatorname{cosec} \lambda \, d\lambda \, (1 - e^2)/(1 - e^2 \cos^2 \lambda).$$

On replacing $(1 - e^2)$ by $(1 - e^2 \cos^2 \lambda - e^2 \sin^2 \lambda)$, the equation becomes:

$$dr/r = \operatorname{cosec} \lambda \, d\lambda - e^2 \sin \lambda \, d\lambda/(1 - e^2 \cos^2 \lambda).$$

The substitution, $e \cos \lambda = \cos u$, transforms this to:

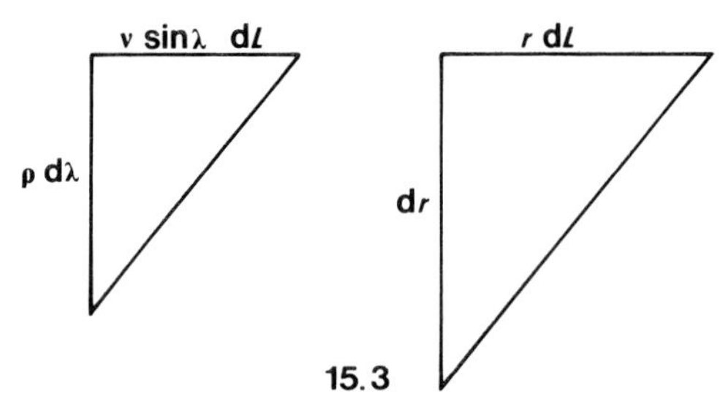

15.3

$$dr/r = \operatorname{cosec} \lambda \, d\lambda - e \operatorname{cosec} u \, du$$

whence the solution is:

$$\ln r = \ln \tan \tfrac{1}{2}\lambda - e \ln \tan \tfrac{1}{2}u + \text{constant},$$

or

$$r = C \tan \tfrac{1}{2}\lambda (\cot \tfrac{1}{2}u)^e.$$

Now the radius of curvature of the spheroid at the pole is a^2/b. By differentiation we find that $dr/d\lambda$ at $\lambda = 0°$ is $\frac{1}{2}C(\cot \frac{1}{2}u_0)^e$ and for true scale at the pole this must be equal to a^2/b. Thus $C = (2a^2/b)(\tan \frac{1}{2}u_0)^e$, so the full formula for r is:

$$(2a^2/b) \tan \tfrac{1}{2}\lambda (\tan \tfrac{1}{2}u_0 / \tan \tfrac{1}{2}u)^e.$$

Identically:

$$\tan^2 \tfrac{1}{2}u = (1 - \cos u)/(1 + \cos u) = (1 - e \cos \lambda)/(1 + e \cos \lambda)$$

so the formula can also be written:

$$r = (2a^2/b) \tan \tfrac{1}{2}\lambda \left(\frac{1 - e}{1 + e} \right)^{\frac{1}{2}e} \left(\frac{1 + e \cos \lambda}{1 - e \cos \lambda} \right)^{\frac{1}{2}e}.$$

It may be noted that if the geometrical projection method illustrated in fig. 13.10 is applied to a spheroid, a reasonable projection for polar regions will be produced, but it will not be exactly orthomorphic. The geometrical construction and the mathematical derivation, when applied to a spheroid, do not necessarily make the same projections.

15.5 Transverse mercator

The projection most widely used nowadays is the transverse mercator, which is orthomorphic and designed to cover meridional zones of longitude extending only a few degrees each side of a central meridian.

There is no geometrical construction for any mercator-type of projection on a sphere or spheroid. For the transverse mercator, formulas have to be derived from the requirements that the projection shall be orthomorphic and symmetrical with respect to a straight central meridian. Since the longitude L is restricted to not more than about $\frac{1}{20}$ radian, it is usual to give formulas as series in powers of L.

The condition of symmetry means that when the sign of L is changed, the easting coordinate must change sign but not value, and the northing must not change. Also, to have true scale on the central meridian, N must be simply m when L is zero (m is the meridian length, section 8.10). Evidently we can write:

$$N = m + AL^2 + BL^4 \ldots$$
$$E = PL + QL^3 + RL^5 \ldots$$

in which the coefficients $A, B \ldots$ and $P, Q \ldots$ depend on ϕ and not on L. To obtain the formulas of these coefficients it is necessary to apply some general theory of conformal representation.

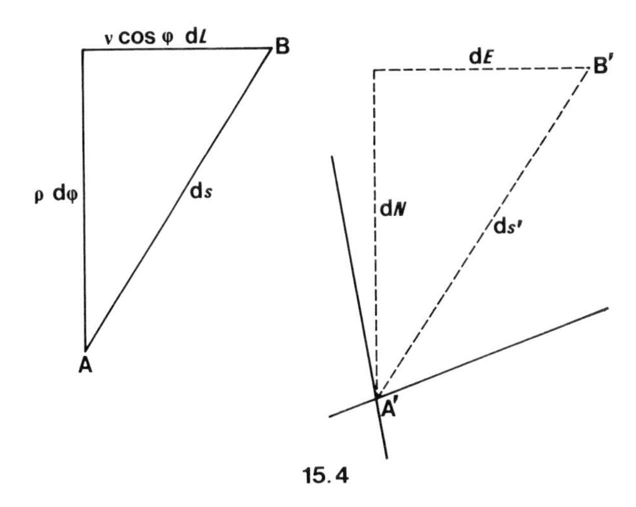

15.4

Let A be the point of latitude ϕ, longitude L. If small changes $\mathrm{d}\phi$ and $\mathrm{d}L$ are made, the linear shifts along meridian and parallel will be $\rho\ \mathrm{d}\phi$ and $\nu \cos\phi\ \mathrm{d}L$ as on fig. 15.4. There will be corresponding changes $\mathrm{d}N$ and $\mathrm{d}E$ in the projection coordinates. Since N and E are functions of two variables, ϕ and L, we can write in the usual differential notation:

$$\mathrm{d}N = \frac{\partial N}{\partial \phi}\,\mathrm{d}\phi + \frac{\partial N}{\partial L}\,\mathrm{d}L$$

and

$$\mathrm{d}E = \frac{\partial E}{\partial \phi}\,\mathrm{d}\phi + \frac{\partial E}{\partial L}\,\mathrm{d}L.$$

Now the 'true' length of AB is given by:

$$(\mathrm{d}s)^2 = \rho^2(\mathrm{d}\phi)^2 + \nu^2 \cos^2\phi\,(\mathrm{d}L)^2$$

and the projection length A′B′ is given by:

$$(\mathrm{d}s')^2 = (\mathrm{d}N)^2 + (\mathrm{d}E)^2.$$

(The two unbroken lines through A′ represent the meridian and parallel, and they are not necessarily at right angles, but of course they will be so if the projection is orthomorphic.) Substituting the formulas for dN and dE, and rearranging, gives:

$$(\mathrm{d}s')^2 = \left[\left(\frac{\partial N}{\partial \phi}\right)^2 + \left(\frac{\partial E}{\partial \phi}\right)^2\right](\mathrm{d}\phi)^2 + 2\left[\left(\frac{\partial N}{\partial \phi}\right)\left(\frac{\partial N}{\partial L}\right) + \left(\frac{\partial E}{\partial \phi}\right)\left(\frac{\partial E}{\partial L}\right)\right](\mathrm{d}\phi)(\mathrm{d}L) + \left[\left(\frac{\partial N}{\partial L}\right)^2 + \left(\frac{\partial E}{\partial L}\right)^2\right](\mathrm{d}L)^2.$$

At this point the requirement of orthomorphism is introduced, that is, that the scale factor ds'/ds is to be independent of the relative values of dϕ and dL. For d$L = 0$, we find that $(\mathrm{d}s'/\mathrm{d}s)^2$ is $[(\partial N/\partial\phi)^2 + (\partial E/\partial\phi)^2]/\rho^2$ along the meridian, and for d$\phi = 0$, it is $[(\partial N/\partial L)^2 + (\partial E/\partial L)^2]/\nu^2 \cos^2\phi$ along the parallel. So these two fractions must be equal. Moreover the full expression for $(\mathrm{d}s'/\mathrm{d}s)^2$ must be equal to these in general, that is for any values of dϕ and dL. Looking back at formulas for $(\mathrm{d}s')^2$ and $(\mathrm{d}s)^2$ it is easily seen that their ratio cannot *in general* be equal to its values in the two special cases, unless the coefficient of (dϕ)(dL) is identically zero. Thus we come to the two 'Cauchy–Riemann' conditions for conformal representation:

$$\nu^2\cos^2\phi\left[\left(\frac{\partial N}{\partial \phi}\right)^2 + \left(\frac{\partial E}{\partial \phi}\right)^2\right] = \rho^2\left[\left(\frac{\partial N}{\partial L}\right)^2 + \left(\frac{\partial E}{\partial L}\right)^2\right]$$

and

$$\frac{\partial N}{\partial \phi}\frac{\partial N}{\partial L} + \frac{\partial E}{\partial \phi}\frac{\partial E}{\partial L} = 0.$$

By differentiating the series for N and E:

$$\frac{\partial N}{\partial \phi} = \rho + \frac{\mathrm{d}A}{\mathrm{d}\phi}L^2 + \frac{\mathrm{d}B}{\mathrm{d}\phi}L^4 \ldots \qquad \frac{\partial N}{\partial L} = 2AL + 4BL^3 \ldots$$

$$\frac{\partial E}{\partial \phi} = \frac{\mathrm{d}P}{\mathrm{d}\phi}L + \frac{\mathrm{d}Q}{\mathrm{d}\phi}L^3 \ldots \qquad \frac{\partial E}{\partial L} = P + 3QL^2 + 5RL^4 \ldots$$

These are substituted in the two conditions, which will be satisfied identically only if the terms not containing L equate, and the coefficient of each power of L is zero. This means some rather cumbersome algebra and some differentiations to get dA/dϕ, etc. For instance, in the first of the Cauchy–Riemann conditions, the terms not containing L give $\rho^2\nu^2\cos^2\phi = \rho^2P^2$, hence $P = \nu\cos\phi$. Then the terms in L in the second condition require $2A\rho + P(\mathrm{d}P/\mathrm{d}\phi) = 0$. It is found that $\mathrm{d}P/\mathrm{d}\phi = -\rho\sin\phi$, whence $A = \frac{1}{2}\nu\sin\phi\cos\phi$. And so on. As usual in this sort of process, the successive coefficients become more and more complicated, but in the higher powers of L some simplification is possible by making use of the fact that e^2 is small, about $\frac{1}{150}$. The formulas as usually published are:

$$N = m + \tfrac{1}{2}\nu \cos\phi \sin\phi\, L^2 + \tfrac{1}{24}\nu \cos^3\phi \sin\phi\, (5 - \tan^2\phi)\, L^4 \ldots$$

$$E = \nu \cos\phi\, L + \tfrac{1}{6}\nu \cos^3\phi(\nu/\rho - \tan^2\phi)\, L^3$$
$$+ \tfrac{1}{120}\nu \cos^5\phi\, (5 - 18\tan^2\phi + \tan^4\phi)\, L^5 \ldots$$

It is worth while to make rough estimates of the likely maximum numerical values of the terms in the series. The radius of curvature ν is about 6.4 million metres. With $L = \frac{1}{20}$ radian, the term in L^5 has a maximum value of about 80 millimetres, so this term will often be negligible, and there is evidently no need to consider the term in L^6.

With $\phi = 50°$ and $L = 2°$, and the spheroid dimensions used before, the coordinates are made up as follows:

N	E
5 540 866.416 m	143 392.026 m
+ 1 917.151	− 5.023
+ 0.288	− 0.006
5 542 783.85 m	143 387.00 m

The N above is referred to an origin on the equator. In practice an origin just south of the mapped area would be adopted. With origin at 49 for instance, the northing would be 113 137.25 m.

The scale factor is the square root of either of the two fractions given above for the special cases, and if the series for the partial differential coefficients are substituted, some algebraic work shows that the factor is $1 + \frac{1}{2}(\nu/\rho)\cos^2\phi\, L^2 \ldots$ It is more convenient to express this in terms of the easting coordinate E which is $\nu \cos\phi\, L \ldots$ and write the factor as $1 + \frac{1}{2}E^2/(\nu\rho) \ldots$ At the point calculated above the factor is 1.000 252 . . . The fourth order term can be calculated, but in practice it is hardly likely to amount to as much as one part per million.

15.6 Reduction factor

In most of the projections that have been described the scale is correct along certain lines and is greater than the nominal scale over the rest of the projection. This leads to the idea that if an overall small reduction of scale is made, the scale factor will be less than 1 in some parts of the projection, and so the differences from the nominal scale can be numerically minimised. The most suitable reduction factor will of course depend on the extent of the area to be covered by the mapping system.

For instance, if E has a maximum value of 300 km, where the above formula shows the factors to be 1.001 105, an overall reduction of 552 parts per million will leave the scale factor 1.000553 at the extremes of the mapped area, and make it 0.999448 on the central meridian. It is then found that the scale is 'true' at $E = 212$ km.

In calculating the coordinates of points in a projection with a reduction factor, a possible procedure is to reduce the linear dimensions of the reference spheroid 'model' and then compute with the original projection formulas. The surveyor working in such a system must remember to apply the reduction factor also to his linear measurements before using them for calculating projection coordinates, (*see* section 16.3).

The Ordnance Survey projection for its published maps has the factor 0.999601271.

15.7 Cassini

In a transverse version of the cylindrical equidistant projection, the easting of a point A is the length (on the model) of a line AD which is perpendicular to a chosen central meridian. But on a spheroid, what line AD is to be used? It could be the geodesic, the shortest route from A to the meridian. Another possibility is to locate D so that A is on the plane vertical section perpendicular to the meridian at D. In any case, given the ϕ and L of point A, it is necessary to calculate the latitude of D and the length of DA; then the northing coordinate is the length of meridian from an adopted origin to the latitude of D.

As this projection is rarely, if ever, used nowadays, the tedious calculations are not set out here. The formulas (for the geodesic case) are:

$$N = m + \tfrac{1}{2}\nu \sin\phi \cos\phi\, L^2 + \tfrac{1}{24}\nu \sin\phi \cos^3\phi(5\nu/\rho - \tan^2\phi)\, L^4 \ldots$$
$$E = \nu \cos\phi\, L - \tfrac{1}{6}\nu \cos\phi \sin^2\phi\, L^3 - \tfrac{1}{120}\nu \cos^3\phi \sin^2\phi(8\nu/\rho - \tan^2\phi)\, L^5 \ldots$$

In this system the coordinates of the point $\phi = 50°$, $L = 2°$ are:

N	E
5540866.416 m	143392.031 m
+ 1917.151	− 17.088
+ 0.288	− 0.002
5542783.86 m	143374.94 m

with origin on the equator. The larger value for the easting in the transverse mercator projection results from the necessary expansion of scale in this direction so as to match the inevitable expansion north–south, to make the transverse mercator orthomorphic.

15.8 Conical

A conical projection representing part of a spheroidal surface is suitable for a mapping series covering an area extending in longitude but of small extent in latitude. This means that the longitude of a point cannot be treated as a small angle, but advantage may be taken of the narrowness of the latitude range. However, the geometry of conical projections is quite simple, because the radii r are functions of latitude only and the projections have a straightforward polar structure; and nowadays the advantage just mentioned may not be needed if a computing facility to 10-figure capacity is available.

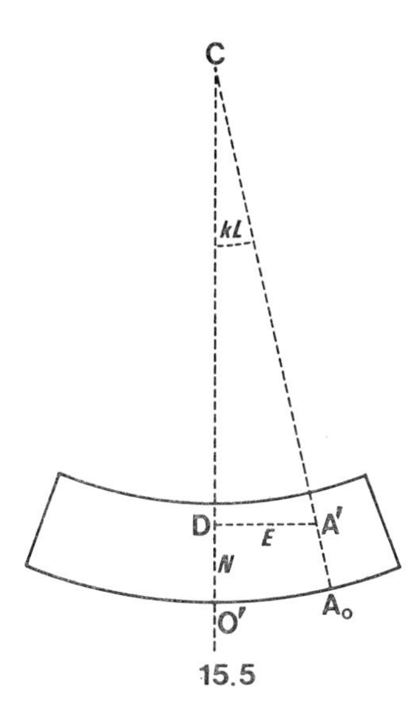

15.5

The basic structure is illustrated in fig. 15.5 on which the curved strip represents the mapped area. It will be convenient to adopt a central meridian as zero of longitude, and as one of the coordinate axes. Point C is the centre for the circular parallels and straight meridians. A point A on the spheroid at latitude ϕ, longitude L is represented by A′, the position of which is given by the angle kL and the length r of the radius CA′. The properties of a particular conical projection are embodied in the function giving r in terms of ϕ. Though (r, kL) are polar coordinates of A′, a plane rectangular system will usually be required for actual plotting operations. The easting of A′ is the length of A′D perpendicular to the central meridian, and the northing is the distance of D from some adopted origin O′. For a particular projection CO′ will be a known length r_0, so the coordinates of A′ are calculated thus:

$$N = r_0 - r\cos kL = (r_0 - r) + 2r\sin^2 \tfrac{1}{2}kL$$
$$E = r\sin kL.$$

Since r_0 and r will be numbers of magnitude comparable with the dimensions of the spheroid, these formulas call for 10-figure capacity to do calculations at 0.001 metre precision.

The required computer capacity can be reduced by the procedure indicated

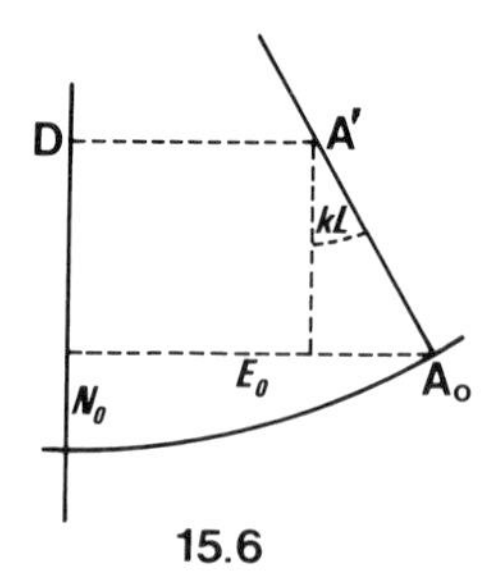

15.6

in fig. 15.6. A list is prepared of coordinates N_0, E_0 of points on the parallel through O′, that is, $E_0 = r_0 \sin kL$, $N_0 = E_0 \tan \frac{1}{2}kL$. Another table may be prepared giving the lengths $A_0A' = (r_0 - r)$ in terms of ϕ calculated from the projection formula. Then, to get the coordinates of A′ we have $E = E_0 - (r_0 - r) \sin kL$ and $N = N_0 + (r_0 - r) \cos kL$. In practice, $(r_0 - r)$ will have maximum values of the order of 250000 metres.

15.9 Conical orthomorphic

Two methods for constructing a conical orthomorphic projection are outlined below.

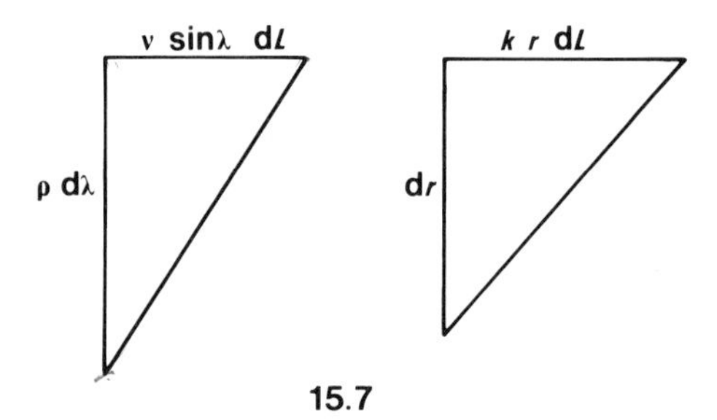

15.7

The elementary triangles, using co-latitude, are shown in fig. 15.7 and the condition for orthomorphism is:

$$\begin{aligned} \mathrm{d}r/rk &= (\rho/\nu) \operatorname{cosec} \lambda \, \mathrm{d}\lambda = (1 - e^2)(1 - e^2 \cos^2 \lambda)^{-1} \operatorname{cosec} \lambda \, \mathrm{d}\lambda \\ &= \operatorname{cosec} \lambda \, \mathrm{d}\lambda - e^2 \sin \lambda \, \mathrm{d}\lambda \, (1 - e^2 \cos^2 \lambda)^{-1}. \end{aligned}$$

This can be dealt with by means of the kind of substitution used before, $\cos w = e \cos \lambda$, and the integral is found to be:

$$\ln r = \text{constant} + k \ln \tan \tfrac{1}{2}\lambda - ek \ln \tan \tfrac{1}{2}w$$

or,

$$r = C(\tan \tfrac{1}{2}\lambda)^k (\tan \tfrac{1}{2}w)^{-ek}.$$

The parallel λ is true to scale if $\nu \sin \lambda = kr$, that is, $\ln r = \ln(\nu \sin \lambda) - \ln k$. Hence, for standard parallels at λ_1 and λ_2, equating formulas for r, we have:

$$\ln(\nu_1 \sin \lambda_1) - \ln k = \ln C + k \ln \tan \tfrac{1}{2}\lambda_1 - ek \ln \tan \tfrac{1}{2}\omega_1$$
$$\ln(\nu_2 \sin \lambda_2) - \ln k = \ln C + k \ln \tan \tfrac{1}{2}\lambda_2 - ek \ln \tan \tfrac{1}{2}\omega_2.$$

By subtraction a formula for k is obtained, and then a formula for C. Plane coordinates can be calculated from the polar system (r kL) if sufficient computer capacity is available.

The parallels λ_1 and λ_2 will be chosen near the limits of the range of latitude, so as to achieve a reasonable balance between negative scale 'errors' between the two standard parallels and positive 'errors' elsewhere. Indeed, it is possible by some more elaborate mathematical working, to choose the standard parallels so that the scale 'error' has a 'minimax' value, that is, its maximum absolute numerical value is as small as possible within the range of latitude to be covered.

Another way to a conical orthomorphic construction starts with the cone touching the spheroid along parallel ϕ_0 in the middle of the latitude range, and then an overall reduction factor is applied so that two parallels become true to scale. Following the construction mentioned in section 13.12, the cone is opened out, the constant k is $\sin \phi_0$ and the radius of the standard parallel ϕ_0 is $r_0 = \nu_0 \cot \phi_0$. The projection radius for latitude ϕ could be expressed in terms of r_0 and $(\phi - \phi_0)$, but it has been found simpler to use the meridian arc length m, which here refers to the length from ϕ_0 to ϕ.

Let the formula be:

$$r = r_0 - Pm - Qm^2 - Rm^3 - \ldots$$

Now, $dm/d\phi = \rho$, so $dr/d\phi = (dr/dm)(dm/d\phi)$
$$= -\rho\,(P + 2Qm + 3Rm^2 + \ldots).$$

Since r decreases as ϕ increases, the condition for orthomorphism is to be written:

$$-dr/(\rho\, d\phi) = kr/(\nu \cos \phi)$$

that is, $(P + 2Qm + 3Rm^2 \ldots)\,\nu \cos \phi = k(r_0 - Pm - Qm^2 \ldots)$.

The $\nu \cos \phi$ must now be expressed as $\nu_0 \cos \phi_0$ + series in powers of m by Taylor's theorem. It starts off:

$$\nu \cos \phi = \nu_0 \cos \phi_0 - m \sin \phi_0 - \tfrac{1}{2}(m^2/\rho_0) \cos \phi_0 \ldots$$

This is substituted in the above equation and coefficients of like powers of m are equated, with the final result:

$$r = r_0 - m - m^3/(6\nu_0\rho_0) - m^4 \tan \phi_0 (1 - 4e^2 \cos^2 \phi_0)/(24\nu_0^2\rho_0).$$
$$- m^5(5 + 3 \tan^2 \phi_0)/(120\nu_0^3\rho_0) \ldots$$

omitting some terms with factor e^2 that are too small to matter in practice.

The scale factor $-dr/dm$ may be taken as:

$$1 + m^2/(2\nu_0\rho_0) + m^3 \tan \phi_0/(6\nu_0^2\rho_0) \ldots$$

and an overall reduction factor is optional. A simple fraction close to $1 - m^2/(4\nu_0\rho_0)$, taking m as its maximum value in the mapped area, might be adopted.

15.10 Transformation to a sphere

A different approach to the construction of projections of a spheroid surface is by way of transforming, that is representing, it on a sphere. Then any plane representation of the sphere will complete the process. This obviously has many possibilities. However, there is not much point in doing it this way unless some desired property is preserved through both stages.

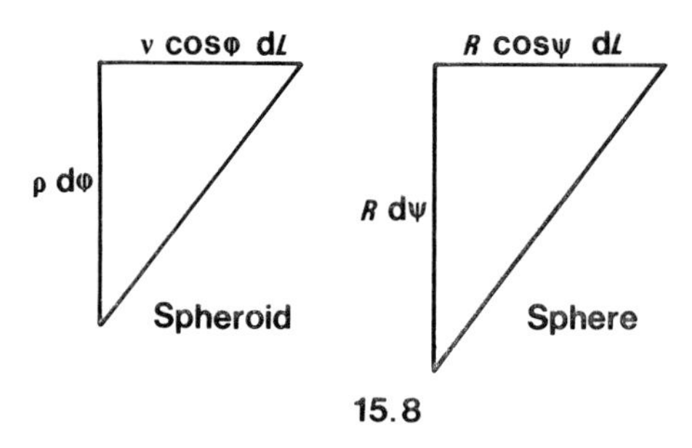

15.8

By use of the elementary triangles shown in fig. 15.8, the point of latitude ϕ, longitude L, on the spheroid is represented by a point of latitude ψ, longitude L, on a sphere of radius R.

For orthomorphic transformation:

$$d\psi/\cos\psi = \rho\, d\phi/(\nu\cos\phi)$$

that is,

$$\sec\psi\, d\psi = \sec\phi\, d\phi(1 - e^2)(1 - e^2\sin^2\phi)^{-1}$$

which can be written as:

$$\sec\psi\, d\psi = \sec\phi\, d\phi - e^2\cos\phi\, d\phi(1 - e^2\sin^2\phi)^{-1}.$$

This has a close resemblance to an equation is section 15.3 and a solution is:

$$\ln\tan(45^\circ + \tfrac{1}{2}\psi) = \ln\tan(45^\circ + \tfrac{1}{2}\phi) - e\ln\tan(45^\circ + \tfrac{1}{2}v)$$

where $\sin v = e\sin\phi$.

As an example take $\phi = 40^\circ$, $e^2 = 0.006\,694\,542$. Then we find that $v = 3.014\,747\,74^\circ$, hence $\psi = 39.810\,693\,16^\circ = 39^\circ\,48'\,38.495''$. There are several ways of expressing the relation between ψ and ϕ, such as the series difference:

$$\psi = \phi - e^2\sin\phi\cos\phi - \tfrac{5}{6}e^4\sin^3\phi\cos\phi$$
$$- \tfrac{1}{30}e^6\sin^3\phi\cos\phi\,(26\sin^2\phi - 5)\ldots$$

The points (ψ, L) may now be represented on a plane by any spherical projection. For instance, a stereographic projection centred at any point on the sphere will result in an orthomorphic projection of the spheroid. It is interesting to note that if the transverse mercator projection of the sphere is made (section 14.1) the result will be a transverse orthomorphic projection of the spheroid, but it will not have exactly the same formula or properties as the transverse mercator described in section 15.5. For instance, the scale will not be uniform along the central meridian.

In transforming orthomorphically to the sphere, its radius R is arbitrary and could, in practice, be chosen to satisfy some particular condition, such as giving average true scale over the mapped area.

CHAPTER 16

Using plane systems

When a survey is connected to points in a national survey framework, the surveyor will probably wish (or be required) to express the positions of the surveyed points in the plane coordinate system. In theory he could calculate spheroidal coordinates of the points and convert to N and E by using the projection formulas. In practice, since spheroidal coordinates may not be required for many, or indeed any, of the points, this two-stage method is usually avoided by means of a procedure in which the survey geometry (angles and lengths) is used so as to give directly the correct plane coordinates of the new points. The procedure must take account of the precise relationship between the spheroid and its plane representation, that is the projection structure.

16.1 Spheroid and plane

The relationship between geometry on the spheroid and its corresponding plane representation is illustrated in fig. 16.1. Through point A on the spheroid the meridian is AP, and AB is a line of length s, that might be a line in a triangulation system or a leg of a traverse; its spheroid bearing at A is A. This geometry appears on the projection as A′P′and A′B′, both of which will in general be curves (but A′P′ will be straight in a conical projection). The projection bearing A' will be equal to A in an orthomorphic projection; the projection length s' will generally differ from s on account of projection scale properties and any scale reduction factor in use.

Lines A′N and A′E are straight grid lines along which one or other of the plane coordinates is equal to the corresponding coordinate of A′. The angle between A′P′ and A′N may be called *convergence*. This word has several meanings and should be defined carefully when used. The angle between

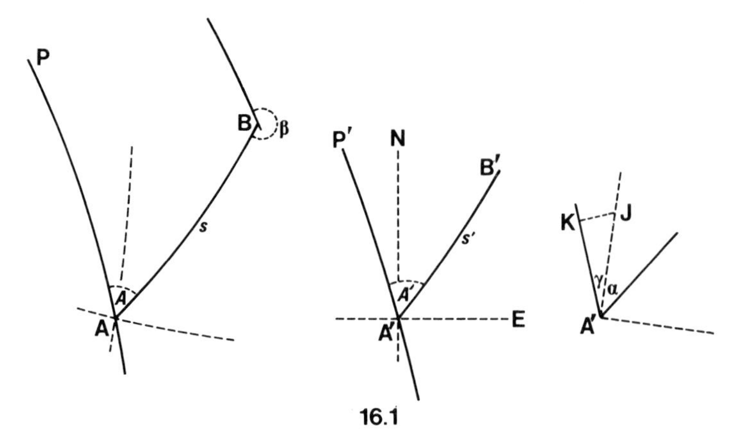

16.1

A′N and A′B′ is the grid bearing of the line A′B′; it is a direction referred to A′N as zero.

Just as lines on the spheroid have their representations in the plane system, so any line on the projection must be representative of some line on the spheroid. In the first diagram in the figure, the broken lines on the spheroid are the lines that appear as A′N and A′E on the plane; they will be at right angles in an orthomorphic projection.

The third diagram in the figure is intended to show an 'infinitesimal' area round the point A on the *spheroid*. The angle marked γ is represented by the 'convergence' angle mentioned above and will be equal to it in an orthomorphic projection. A method for calculating γ will be required.

Let J be a point on the spheroid grid line and let JK be perpendicular to the meridian. Let the differences of latitude and longitude between A and J be $\mathrm{d}\phi$ and $\mathrm{d}L$. Then the linear lengths are AK $= \rho\,\mathrm{d}\phi$, KJ $= \nu\cos\phi\,\mathrm{d}L$. There is no change of easting coordinate from A to J and this means that, referring to a formula in section 15.5:

$$\mathrm{d}E = \frac{\partial E}{\partial \phi}\,\mathrm{d}\phi + \frac{\partial E}{\partial L}\,\mathrm{d}L = 0.$$

Now $\tan\gamma = \mathrm{JK/KA} = (\nu\cos\phi\,\mathrm{d}L)/(\rho\,\mathrm{d}\phi)$, and by the above equation:

$$\frac{\mathrm{d}L}{\mathrm{d}\phi} = -\left(\frac{\partial E}{\partial \phi}\right)\bigg/\left(\frac{\partial E}{\partial L}\right).$$

Hence,

$$\tan\gamma = -\frac{\nu\cos\phi}{\rho}\left(\frac{\partial E}{\partial \phi}\right)\bigg/\left(\frac{\partial E}{\partial L}\right).$$

16.2 Projection coordinate differences

It is reasonable to assume that in order to calculate projection coordinates directly from survey geometry, some formulas could be found to give differences of coordinates from point to point in terms of s and the 'true' bearing A. This can be done, but it is found to be simpler and more convenient to use the angle marked α, equal to $A - \gamma$, instead of A. That this is so obviously due to the fact that plane coordinates are more closely associated with grid lines than with meridians and parallels. One would expect the differences of coordinates to be very nearly $s \cos \alpha$ and $s \sin \alpha$.

This method will be developed in further detail with reference to the transverse mercator projection. In this system $\gamma = L \sin \phi + \frac{1}{3}L^3 \sin \phi \cos^2 \phi \ldots$ and in practice the next term is quite negligible. By substituting $\alpha + \gamma$ for A in formulas for differences of spheroid coordinates, and using the formulas for the transverse mercator projection, and a lot of tedious algebra, the results are:

$$\Delta N = s \cos \alpha + \tfrac{1}{2}E^2 s \cos \alpha/\nu\rho + \tfrac{1}{2}s^2 \sin 2\alpha\,(E + \tfrac{1}{3}s \sin \alpha)/\nu\rho \ldots$$

$$\Delta E = s \sin \alpha + \tfrac{1}{2}E^2 s \sin \alpha/\nu\rho - \tfrac{1}{2}s^2 \cos 2\alpha\,(E + \tfrac{1}{3}s \sin \alpha)/\nu\rho \ldots$$

Note that $E + \frac{1}{3}s \sin \alpha$ is practically the easting coordinate of the point one-third of the way from A to B. After the principal terms $s \cos \alpha$ and $s \sin \alpha$, the others are much smaller, and only approximate values of ν and ρ are needed.

In order to continue this method of computation to further points, it is necessary, as in the calculation of latitudes and longitudes, to get a 'reverse bearing' referred to the spheroid grid line at B – the angle marked β in the figure – so that survey data can be used to find the 'α' for the next line. The formula for this is:

$$\beta = \alpha \pm 180° - s \cos \alpha\,(E + \tfrac{1}{2}s \sin \alpha)/\nu\rho$$

in which the last term is in radian.

For a numerical example consider the two points A and B of section 10.7. Take A to be at longitude 2° east of the central meridian, so that the longitude of B is then 2° 24′ 07.566″. Coordinates calculated by transverse mercator formulas (section 15.5) with origin on the equator, are:

Point A Lat. 50° 38′ 51.208″ Long. 2° 00′ 00.000″

$$N = 5\,612\,897.956 + 1\,909.092 + 0.274 = 5\,614\,807.322 \text{ m}$$
$$E = 141\,456.816 - 5.595 - 0.006 = 141\,451.215 \text{ m.}$$

Point B Lat. 50° 26′ 43.362″ Long. 2° 24′ 07.566″

$$N = 5\,590\,407.515 + 2\,757.691 + 0.579 = 5\,593\,165.785 \text{ m}$$
$$E = 170\,624.884 - 9.388 - 0.015 = 170\,615.481 \text{ m.}$$

At point A we find that $\gamma = 5\,567.47'' + 0.91'' = 1°\,32'\,48.38''$ so that α is $128°\,07'\,19.1'' - 1°\,32'\,48.4'' = 126°\,34'\,30.7''$; and with $s = 36\,305.92$ m, the differences of coordinates are:

$$\Delta N = -\,21\,633.872 - 5.320 - 2.344 = -\,21\,641.536 \text{ m}$$
$$\Delta E = +\,29\,156.396 + 7.161 + 0.709 = +\,29\,164.266 \text{ m.}$$

These make the coordinates of B to be 5593165.786 and 170615.481, in good agreement with the values calculated directly from the latitude and longitude.

From the bearing formula we get:

$$\beta = 126^\circ\, 34'\, 30.7'' + 180^\circ + 17.1'' = 306^\circ\, 34'\, 47.8''.$$

But the 'γ' angle at B is $6667.43'' + 1.58'' = 1^\circ 51'\, 09.0''$, so β is $308^\circ\, 25' 56.8'' - 1^\circ\, 51'\, 09.0'' = 306^\circ\, 34'\, 47.8''$, as obtained above.

It is to be noted that in the method of computation described above the 'data' to go into the calculations are the spheroidal angles and lengths. Projection properties are incorporated in the small terms in the formulas for ΔN and ΔE. In each of these formulas, the first two terms taken together contain the scale factor $1 + \frac{1}{2}E^2/\nu\rho$. Another method of approach is described in the next section.

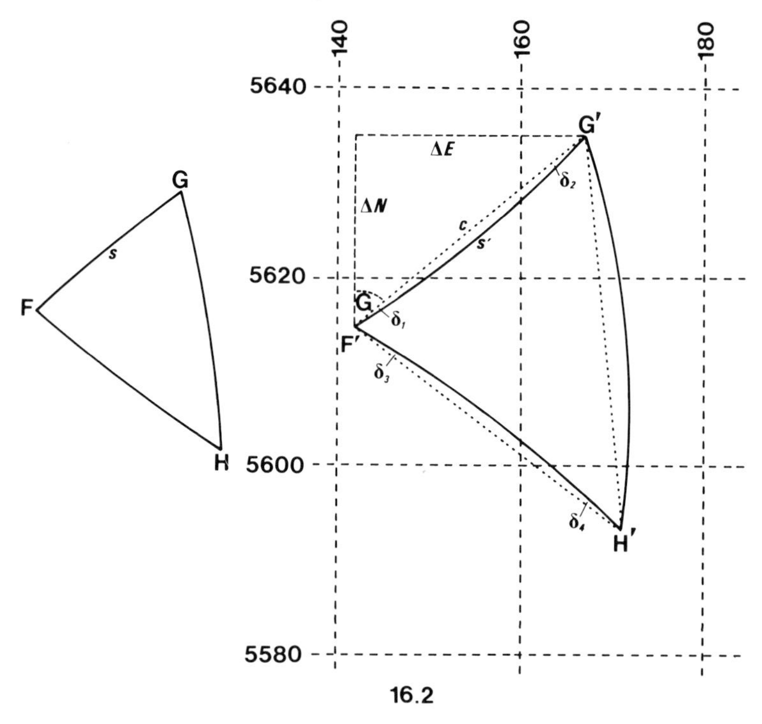

16.2

16.3 Working in projection coordinates

Consider in detail the relationship between a triangle FGH on the spheroid, and its representation F′G′H′ on the projection system indicated by the square grid in fig. 16.2. In diagrams like this the curvatures have to be exceedingly exaggerated to show them: in transverse projections the curves are concave towards the central meridian. The *straight* chords joining F′G′H′ are also shown. If the dimensions of the chord triangles can be derived from the survey system, the coordinates of points in the projection system can be calculated by plane trigonometry.

Suppose that the plane coordinates of F′ are known. Then the coordinates of G′ could be calculated by simple plane trigonometry from the *chord* length c and the grid bearing marked G.

In an orthomorphic projection the grid bearing of the *curve* F′G′ at F′ is the same as the angle α of section 16.2. Hence if the small angle marked δ_1 is found, the grid bearing G is $\alpha - \delta_1$. In practice the angles like δ_1 amount to a few seconds; they have been called 'arc–chord' angles, or 't − T' corrections.

The curve length marked s' differs from the 'true' length s in accordance with the projection scale factor as well as any general reduction factor that is in use. The difference between s' and c is quite negligible in all practical cases. For the shorter lines, the scale factor can be taken as that appropriate to the middle of the line.

To obtain angle δ_1 we note that $\tan G = \Delta E/\Delta N$ and $\tan \alpha = \mathrm{d}E/\mathrm{d}N$ when s is zero. Using the formulas for ΔN and ΔE it can be shown that:

$$\delta_1 = \alpha - G = \tfrac{1}{2}s \cos \alpha \,(E + \tfrac{1}{3}s \sin \alpha)/\nu\rho \text{ radian}$$

and then finally $\Delta N = c \cos G, \;\; \Delta E = c \sin G$.

As before, a reverse bearing is required for continuing to the next line. Of course, the reverse bearing of the chord is simply $G \pm 180°$. By applying (in the case illustrated, subtracting) the angle marked δ_2, the grid bearing of the *curve* G′F′ is obtained. Then angles in the survey system can be added or subtracted to get the forward grid bearing of the next line. And so on. The formula for δ_2 is :

$$\tfrac{1}{3}s \cos \alpha \,(E + \tfrac{2}{3}s \sin \alpha)/\nu\rho \text{ radian.}$$

For $s \cos \alpha$, ΔN may be used.

Since the scale factor is not a linear function of the easting, long lines should be divided into sections for the calculation of conversions to chord length. The line used in section 16.2 and shown as F′H′ on fig. 16.2 will be divided into three parts: the mean eastings for the three parts are approximately 146 312, 156 033 and 165 754 m, and the corresponding scale factors are 1.000 262 8, 1.000 298 9 and 1.000 337 3. The additions to the true length are 3.180, 3.617, 4.082 m making the chord length 36 316.799 m. The angle δ_3 is found to be 8.3″ so the grid bearing of the chord is 126° 34′ 39.0″. By plane trigonometry $\Delta N = -21\,641.528$ m and $\Delta E = +29\,164.262$ m. The arc–chord angle δ_4 is 8.8″, so the reverse grid bearing of the curve H′F′ is 306° 34′ 39.0″ + 8.8″ = 306° 34′ 47.8″, in agreement with the other computation.

The discrepancies of a few millimetres from the previous values of ΔN and ΔE could be considered negligible; but they would be smaller by dividing the line into more than three parts. However, the method just described will be used mostly for calculating coordinates in traverses, with lines very much shorter than 36 km.

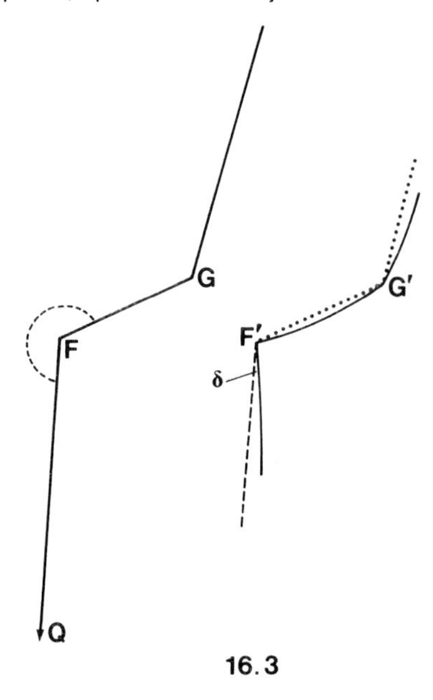

16.3

16.4 Traverse computation

Fig. 16.3 illustrates the problem of starting a traverse computation in a coordinate system. In this case FG is the first line of the traverse, and to obtain a starting bearing the surveyor sights another coordinated point Q and measures the angle indicated. He requires the grid bearing and projection length of the *chord* (dotted line) F′G′. From the known plane coordinates of F′ and Q′, the grid bearing of the chord F′Q′ is calculated by plane trigonometry. The angle δ is calculated and (in the case illustrated) subtracted to give the grid bearing of the *curve* F′Q′. Assuming an orthomorphic projection, the measured angle can then be added to give the grid bearing of the *curve* F′G′. The calculations then continue as in the previous section.

It will be seen that if the traverse continues in the same general direction, the 'δ' angles will always have to be subtracted. Hence, although these angles are very small, their effects on the grid bearings will accumulate, and if they are neglected a false and large misclosure of bearing may be found at the end of the traverse.

16.5 Adjustment on the plane

Refer back to fig. 16.2 and suppose that GFH is part of a triangulation system. Let F_0 be the *observed* value of angle GFH: then $F_0 + \delta_1 + \delta_3$ could be regarded as the 'observed' value of the angle G′F′H′ of the *chord triangle.* Similarly for the other observed angles. Thus it is possible to see the triangulation as made up of plane rectilinear triangles on the projection plane: the survey measurements are to be transformed to their 'flat–Earth' equivalents.

This plane triangulation contains the errors of observation and will require adjustment as usual. The conditions for adjustment will be mathematically the same as for adjustment on the spheroid, the main difference being that the three angles of each triangle must total exactly 180°. The spherical excess is absorbed into the arc–chord angles.

Any measured lengths can also be included in the adjustment, after they have been converted to 'projection lengths' by the application of scale factors and any general reduction factor.

It will of course be necessary to know the approximate positions of the points in the coordinate system, so that the small arc–chord angles and linear modifications can be made with sufficient accuracy. This may require some preliminary approximate calculations; but if an accurately drawn working diagram of the survey is kept, it should be possible to read off the information for calculating these small quantities.

After adjustment on the plane, final coordinates are calculated by plane trigonometry.

16.6 The Hotine method

When the projection is not orthomorphic, the use of the methods described in previous sections will require small changes to be made to observed angles, in addition to the arc–chord and scale 'corrections'. To avoid this further complication, M. Hotine suggested that within a small area, the coordinate system could be made almost orthomorphic by locally 'compressing' or 'expanding' it in one direction so that the scale becomes the same in all directions. Then a survey could be calculated using 'true' angles, and the resulting coordinates returned to the proper projection system by reversing the compression or expansion.

For instance, the Cassini projection has true scale along the east–west grid lines and exaggeration by the factor $1 + \frac{1}{2}E^2/\nu\rho$ in the perpendicular direction. Over a small area, projection coordinates could be referred to a temporary local origin within the area and the northings proportionally reduced to cancel the scale exaggeration. Then the new survey points could be calculated using unmodified angles, and their local northings then expanded by the scale factor before being returned from the local origin to the general system.

Examples 2

(Assume spheroid dimensions $a = 6\,378\,160$ m, $f = 1/298.25$)

1. At what latitude on the spheroid is ρ equal to (i) a, (ii) b?
2. At what latitude on the spheroid is the parallel equal to exactly half the length of the equator?
3. Calculate the length along the meridian from latitude 30° to latitude 50°, also the length of the direct chord distance between these points.
4. At point A, spheroid latitude 29° 15′ 38.204″ N, longitude 76 02′ 58.339″ E, the spheroid bearing and distance to point B are 307° 28′ 11.5″ and 18 207.94 m. Calculate the spheroid coordinates of B and the reverse bearing.
5. Point A is 611 metres above the spheroid on the normal at latitude 57° 15′ 32.7″ N, longitude 7° 17′ 39.1″ W, and point B is 18 metres above the spheroid on the normal at latitude 52° 24′ 56.5″ N, longitude 1° 43′ 03.2″ E. Calculate the rectangular coordinates of these points in the system having the Z axis on the spheroid axis, the X axis in the equator at zero longitude and the Y axis at 90° E. Calculate also the length and the direction cosines of the straight line from A to B.
6. A cylindrical projection in normal aspect has the formulas $E = RL$, $N = 2R \tan \frac{1}{2}\phi$. Find the projection bearing and the scale factor on a line of bearing 45° at latitude 45°.
7. A conical projection in normal aspect has the constant of the cone equal to $\frac{2}{3}$, and all the parallels are true to scale. Find the formulas for the projection radii of the parallels and the scale factor along the meridian. At what latitude is the meridian scale true?
8. A secant conical projection (section 13.16) is constructed on a cone that cuts the sphere along parallels 30° and 60°. The positions of the other parallels are found by direct projection from the centre of the sphere.

Answer as for the previous question.

9. An oblique mercator projection of a sphere is to be constructed with the 'projection equator' being a great circle cutting the true equator at 45°. Find formulas for the projection latitude and longitude ϕ', L' in terms of the terrestrial latitude and longitude ϕ, L.

References

MacRobert,T. M. and Arthur, W. (1938) *Trigonometry Part IV*. London: Methuen.

Robbins, A. R. (1976) 'Field and geodetic astronomy'. *Military Engineering* **XIII**, (IX): Ministry of Defence.

Steers, J. A. (1970) *An Introduction to the Study of Map Projections.* University of London Press.

General index

(The numbers refer to sections)

Answers

Examples 1

1. 40.644074°, 56.863412°, 93.679644° Ratios: 0.767627
2. 21.824630°, 38.251037°, 120.037817° Ratios: 0.140776
3. Sides: 15.986539°, 20.736585°, 25.578809°
   Angles: 39.613249°, 55.055610°, 88.270936°
   Ratios: 0.431949
4. $\sin \frac{1}{2}A = \frac{1}{2} \sec \frac{1}{2}\alpha$, $\sin \lambda = 2 \sin \frac{1}{2}\alpha / 3^{\frac{1}{2}}$
5. 33.189642°, 50.728701°, 71.467819° Ratios: 1.094824
6. Two solutions:
   $\gamma = 94.638751°$, $B = 37.761244°$, $C = 135.187371°$
   or $\gamma = 17.981114°$, $B = 142.238756°$, $C = 12.608401°$
7. $\beta = 45°$, $\gamma = 63.101049°$, $C = 116.898951°$
8. $\beta = 67.792346° = (180° - \gamma)$, $A = 41.409622°$
9. Hour angle: 47.674107° Azimuth: 110.207319°
10. Altitude: 30° 50′ 57″ Hour angle: 69° 53′ 38″ = 4h 39m 34.6s
11. Hour angle: 2h 57m 33.1s Altitude: 30° 40′ 31″ Azimuth: 328° 09′ 45‴
12. 232° 42′ 09″
13. $\mathrm{d}B = -\,\mathrm{d}A \sin B \cos C / \sin A$, $\mathrm{d}C = -\,\mathrm{d}A \sin C \cos B / \sin A$
    Error: 13″.

Examples 2

1. (i) 54° 46′ 51.5″ (ii) 35° 18′ 35.1″
2. 60° 04′ 59.7″
3. 2220741.5 m, 2209483.9 m
4. Latitude: 29° 21′ 37.664″ Longitude: 75° 54′ 02.555″ E
   Bearing: 127° 23′ 49.2″

5. *Point* A

   $X = +3\,430\,115$ m
   $Y = -439\,055$
   $Z = +5\,342\,092$

   Point B

   $X = +3\,896\,682$ m
   $Y = +116\,846$
   $Z = +5\,031\,180$

   Length: 789 542 m
   Direction cosines: +0.590 934, +0.704 080, −0.393 788
6. 50° 21′ 39″, 1.2986
7. $r = (3/2)\,R \sin\lambda$: Scale factor: $(3/2)\cos\lambda$: Latitude: 41.81°
8. $r = 2R \cos 15° \sin\lambda\,(\sin\lambda + \cos\lambda)^{-1}$
   Scale factor on meridian: $2 \cos 15°\,(1 + \sin 2\lambda)^{-1}$
   Scale factor on parallel: $2^{\frac{1}{2}} \cos 15°\,(\sin\lambda + \cos\lambda)^{-1}$
   Latitude: 34° 21′ 46″ and 55° 38′ 14″
9. $2^{\frac{1}{2}} \sin\phi' = \sin\phi + \cos\phi \sin L$
   $2^{\frac{1}{2}} \tan L' = \tan L + \tan\phi \sec L$
   or equivalent formulas.